Methodological Approaches to STEM Education Research Volume 4

Methodological Approaches to STEM Education Research Volume 4

Edited by

Peta J. White,

Russell Tytler,

Joseph Paul Ferguson

and John Cripps Clark

Cambridge
Scholars
Publishing

Methodological Approaches to STEM Education Research Volume 4

Edited by Peta J. White, Russell Tytler, Joseph Paul Ferguson
and John Cripps Clark

This book first published 2023

Cambridge Scholars Publishing

Lady Stephenson Library, Newcastle upon Tyne, NE6 2PA, UK

British Library Cataloguing in Publication Data
A catalogue record for this book is available from the British Library

ISBN (10): 1-5275-2589-9
ISBN (13): 978-1-5275-2589-4

CONTENTS

CONTRIBUTORS

Robin Bellingham
School of Education, Deakin University, Australia

Jill P. Brown
School of Education, Deakin University, Australia

John Cripps Clark
School of Education, Deakin University, Australia

Joseph Paul Ferguson
School of Education, Deakin University, Australia

Shelley Hannigan
School of Education, Deakin University, Australia

Margaret Jakovac
School of Education, Deakin University, Australia

Scott Jukes
Institute of Education, Arts and Community, Federation University, Australia

Amrita Kamath
School of Education, Deakin University, Australia

Gahyoung Kim
Faculty of Education, Monash University, Australia

Melinda Kirk
School of Education, Deakin University, Australia

Kongju Mun
Dongduk Women's University, Liberal Arts College, Korea

Amanda Peters
School of Education, Deakin University, Australia

Jo Raphael
School of Education, Deakin University, Australia

Peter Reason
School of Management, University of Bath, United Kingdom

Carly Sawatzki
School of Education, Deakin University, Australia

Shefali Sharma-Wallis
School of Education, Deakin University, Australia

Gary Simpson
Woodleigh School, Victoria, Australia

Cassandra Tytler
Centre for People, Place & Planet, Edith Cowan University, Australia

Russell Tytler
School of Education, Deakin University, Australia

Peta J. White
School of Education, Deakin University, Australia

Sandra Wooltorton
Nulungu Research Institute, University of Notre Dame Australia, Australia

FOREWORD

ANN OSMAN

Being asked to write the foreword to *Methodological Approaches to STEM Education Research Volume 4* was an unexpected honour. It provided me with the opportunity to acknowledge and reflect on the impact education research has had (and continues to have) on shaping and reshaping my ideas, opinions, and understanding of education generally and STEM education specifically. Education research provides the vehicle through which current practices can be monitored and reviewed and new initiatives presented and evaluated to ensure the best possible outcomes for students, their communities and educators.

For me education research tells a story and in so doing shines a light on methodological approaches and practices; presents new directions and approaches to undertaking research where different types of questions are asked and the answers provide new understanding or insights; and identifies issues of concern and provides opportunities for communication and collaboration between educators.

Throughout my career I have been a consumer of education research as I built my understanding of the use of methodological approaches to shine a light on issues in education. My first introduction to the possibilities and outcomes from published research was as a secondary school teacher implementing new curriculum and assessment policies and initiatives that had been developed at a state-wide system level and presented to teachers like me to implement. Later in my career as a bureaucrat I commissioned education research to identify possible solutions to an existing issue, update current policies and suggest future directions or emerging trends in school education.

Without doubt it was as a PhD student that I most experienced the impact and power of education research. When discussing my proposal to use a narrative inquiry approach to present the findings from my investigation of STEM education with my supervisors (late Professor David Clarke and Professor Jan van Driel) challenged my understanding of methodological approaches by asking 'Will you just be telling another good story about STEM education, or will you be presenting actual education research?' Whilst I thought I understood what was and wasn't education research, I had not considered how to convince potential readers (or consumers) of my work that it was education research. By sharing my research I was sharing a narrative that wove together published literature with the findings of the analysis of data collected.

My response to David and Jan was to diligently access (consume) and critique a breadth of published literature and research and to gather, analyse, and synthesise data from different sources whilst constantly checking that I was presenting a narrative that used the data to tell a story worth reading and that would contribute to education research. Being challenged to maintain the quality of education research and its outcomes is as important when completing a PhD as it is when commissioning and consuming research–even more so in times of fiscal constraints where the amount of funding available continues to dwindle.

It is in this environment that the value of being able to attend CAR symposia is so important as it continues to provide a unique space where researchers from novice to experienced can (as I have) present new ideas and approaches to research, challenge existing understandings, present examples of data collected and emergent trends, or identify potential conflicts between data sets. As such it became a safe sounding board for me where, by sharing what I was doing I was able to seek guidance, input, or reassurance from the audience that was collegiate and understood the doubts and insecurities so often felt by students on their journey to a PhD.

Education research is not without its with challenges or tensions. Challenges, for example, conceptualising an idea into a research proposal with the most suitable methodological approach and theoretical framework, or collecting data during a pandemic, or maintaining contact with participants over an extended period can be frustrating and sometimes impact on research outcomes. Tensions, for example, interrogating the data for trends which seem elusive or contradictory to expectations, or accommodating responses from participants that throw a spotlight onto how the research was conceived and how this relates to the researcher's existing position (or stance), require time for reflection and re-thinking. During my PhD I had experienced firsthand the challenges and tensions of education research when using narrative inquiry where I was both the researcher and a participant in the research but fortunately with the guidance of David and Jan who helped me navigate my way.

How the education research is carried out, how quality is maintained, what choices are made and why, matter to both the researcher and the consumer of the research. These choices then become the underpinning pillars or the foundations of the research. This allows the researcher to present, as trustworthy and reliable, the findings (outcomes) for consideration and feedback. Each chapter in Volume 4 provides a unique perspective and offers insights into specific areas of education research. Across the thirteen chapters shared themes emerge including facing unexpected challenges and tensions, developing and applying innovative practices, drawing on and expanding past research and the roles and responsibilities of the researcher in ensuring quality research. In the first five chapters the complex and challenging nature of relationships between researchers and participants and their role in education research is discussed. These chapters also highlight the selection of a methodological approach that provides the flexibility to respond to unexpected changes enabling the research to continue.

Challenges in education research often arise when a methodological approach is applied innovatively as shown by Peters (Chapter 1). She explains the importance of gaining the confidence and trust of participants when using provocations designed to challenge their opinions and encourage them to share ideas. The use of Arts (drama) based strategies as provocations with a group of educators illustrates an innovative approach to exploring and analysing current STEM education policy and to inform future policy.

As discussed by Tytler, White & Sharma-Wallis (Chapter 2), challenges surfaced from multiple contextual restraints when refining and validating science inquiry assessment tasks in their investigation. The successful management of these challenges was made possible by the application of a detailed methodology and careful data management. In so doing the quality and credibility of the research is maintained.

Design-based research (DBR) has much to offer as Kamath and Simpson (Chapter 3), working as co-researchers (researcher and teacher respectively), illustrated by its use to investigate the effectiveness of guided inquiry in senior biology. The challenges encountered began with the clarification of the roles and expectations of the participating teachers. Next was navigating teachers' concerns related to researching in senior secondary classes. Lastly, were the methodological tensions related to the interplay between theory and practice. If not considered there would likely be an impact on the outcomes of the investigation.

The flexibility of design-based research is also valuable. Brown and Sawatzki (Chapter 4) demonstrate this in their DBR and how it and experimenting with using digital technologies enabled the successful shift (due to the once in a hundred-year pandemic) from an in-person to an online program for out-of-field mathematics teachers. This shift resulted in unexpected opportunities to tailor the program to provide activities that could be replicated by the teachers in their classrooms to build their knowledge, understanding, and confidence in teaching mathematics.

Partnerships between universities and schools take many forms including those involving shared research. Kirk & Ferguson (Chapter 5) suggest that university and school research partnerships should bring together research and practice through the development of an aligned community of practice. Using DBR they illustrate how a partnership of educators (from classroom, school, and university) became a community of inquiry working collaboratively and focused on supporting teachers to develop pedagogies that foster primary school students' creativity and critical thinking in science.

The next four chapters offer different perspectives on selecting the methodology most appropriate to the research being undertaken. These include reviewing and then refining a long-established methodological approach, and seeking out new (emerging) methodologies and innovation by combining both traditional and non-traditional approaches.

As Ferguson & Cripps Clark (Chapter 6) discuss, it is often a struggle to realise a new perspective of old problems and suggest that one approach may revisit the past and re-examine how, for example, a scientist (Charles Peirce) wrote about science education. This approach suggests there is value in revisiting past research leading to revising how research is conceived and enacted to have positive impacts on teaching.

With a focus on including financial literacy in school education, research in how teachers become trained in this area by Sawatzki and Brown (Chapter 7) is timely. They discuss an alternate vision of financial education and the use of innovative and non-traditional methodologies (*enterprising research methodologies*) to upskill teachers as financial educators. The program had input from stakeholders and was offered via a free market tertiary education hub. The tensions of delivering a commercial product in an academic environment and how research is contextualised, and the role and identity of the researchers is explored.

Innovation in methodological approaches does not have to mean new or different as Kim & Mun (Chapter 8) show when discussing how innovation can come from re-examining an existing approach. Their investigation shows how Issue-Concept Maps (ICM), used to explore and assess socio scientific issues, could be adapted and used as a tool to construct lesson content, learning strategies, and alternatives to test-based assessment practices.

An exploration of theory and practice can also lead to innovation in education research. Jakovac (Chapter 9) does this by taking an approach to using the theoretical lens of identity and practice to investigate success stories of out-of-field STEM teachers. Once the challenges of finding participants was addressed, the emerging online ethnographic method (netnography) was used to capture the collective professional voice from a group of these teachers through their stories as published in designated Facebook groups. This rich data set provided a better understanding of how identity and practice is shaped.

In the final four chapters the researchers demonstrate the use of new methodological directions in education research and discuss how these could be applied in future investigations of issues for example, continuous engagement with learning and environmental education. The authors also suggest how these new approaches may provide epistemological approaches that allow us to rethink our relations with the human and non-human world.

A different approach to re-purposing the use and application of cooperative inquiry in education research is taken by Wooltorton & Reason (Chapter 10). They begin with a literature review to suggest a rethinking and extension of the use, content, and context of cooperative inquiry where it becomes a collaborative approach to living life as an inquiry. They argue this better fits the current learning and education contexts of enabling life-long learning and re-engagement with learning.

As concerns about the state and future of our environment intensify, White, Raphael, Hannigan & Bellingham (Chapter 11) provide insight into how re-imagining pedagogical practices could provide strategies to explore why (and how) change is needed to how humans impact on our planet Earth. Using technology (Padlet), a collaborative practice was encouraged where researchers *'attuned with'* and engaged with various beings, resulting in changes to pedagogical practices.

Cassandra Tytler (Chapter 12) encourages researchers to think broadly about the possibilities for education research by using an art project undertaken in a public space as an example of place-based methodology. The agency of the artist/researcher is explored to demonstrate the innovative and creative use of art to explore new ways of experiencing our environment and the complexities of interactions within it, much like Peters (Chapter 1) in using drama with STEM education policy.

The use of provocations (thinking with things) to challenge existing practices in education research provides an innovative approach for investigating current issues (for example STEM or environmental education). Jukes (Chapter 13) argues that when existing research practices are questioned unexpected possibilities emerge. The challenge is to then trust these, ensuring the research and research data will be innovative and allow a richer picture of environmental inquiry to emerge.

When writing about the resolution to challenges and tensions faced, it is important that researchers maintain a methodological focus. This will enable them to diligently and honestly explain how the resolution shaped or refined the outcomes of their research, their identity and beliefs, and contribute to the quality of their research. This is evident in Volume 4 as it builds on the principles and practices of methodological approaches in STEM education research first discussed in Volume 1 and continued in Volumes 2 and 3 of the CAR series, and is so doing provides insights into what quality education research looks like. The CAR series continues to encourage education researchers to review existing methodological approaches to refine, refresh, and renew to better fit with contexts being investigated and to develop new practices to better explore current issues.

As someone very interested and invested in STEM education and education research, I commend the contributors to Volume 4 for the exciting applications and future developments in methodological approaches in education they have suggested which I look forward to reading about in the next CAR volumes.

INTRODUCTION

The way we frame research never stands still. In education, as in any discipline, our perspectives on our role as researchers inevitably changes as educational purposes are re-examined in the light of wider shifts in public sensibilities and commitments. An increasing emphasis on social justice is reflected in new interests in participatory research and communities of inquiry. New theories emerge that have implications for the way we approach research and create and view data. Shifting epistemological insights and commitments have repercussions for the status of different types of evidence, and this can be different for diverse research consumers such as government, system policy influencers, administrators, teachers, and other researchers. The 'gold standard' debate in the US is still alive and kicking in public discourse and research policy initiatives. We ignore these wider research methods controversies at our peril.

This volume, the fourth in a series based on the Symposium 'Contemporary Approaches to Research (CAR) in Mathematics, Science, Health and Environmental Education' (https://deakinsteme.org/event-category/car-symposiums/), takes this focus on the need to continually refresh our methodological reach and focus as its primary rationale. The CAR symposium has been running at Deakin more or less continually since the mid 1990s and has been responsible for ongoing interest in research methodologies amongst the STEME (Science, Technology, Engineering Mathematics and Environmental Education) Education Research group, and in a wider set of education researchers who have followed the event over the years. The STEME group is a strong part of research in Education at Deakin University that regularly lists as one of the highest-ranking education research concentrations globally. This methodological focus we argue is a key feature of regenerating research cultures.

In the early days of CAR, there was considerable interest in challenging positivist framings of research data, for instance from interviews or discourse analysis, informed by constructivist and radical constructivist theoretical framings. This interest in epistemology has seen further shifts over the years towards post-structuralist considerations of power and discourse, perspectives on the epistemological bases of classroom video analysis, the role of theory in framing data generation and analysis, and consideration of the conditions for trustworthiness of qualitative or mixed method studies. More recently, we have seen greater representation of Design Based Research, reflecting critical research perspectives that privilege transformation, of arts-based methodologies, and of post-qualitative inquiry allied with challenges to 'business as usual' in the conduct of education in the face of the Anthropocene crises.

This volume is no different in its exploration of contemporary issues, and in the variety of methodologies being canvassed. Unlike most methodological texts, the CAR presentations focus around particular methodological contexts to explore challenges or questions that are grounded in a way that is often papered over in publications focused on findings, or in texts focused on particular methodological framings. The chapters presented extend well beyond the presentation resulting in this publication being, in many ways, a 'user's guide' to methodology in education research.

The first set of chapters in this volume deals with a variety of forms of participatory classroom research, with methodological foci canvassing ways of gaining the trust of participants, of flexibly adjusting Design Based Research processes to be sensitive to teachers' needs, of teachers engaging with the role of co-researcher, of flexibility to changing context of COVID-19 and the unexpected affordances of this pivot, and finally how a community of inquiry theoretical framing can inform participatory processes. Across these chapters we can see the wider variety of settings and changing contexts, and theoretical work that can usefully inform such participatory research. We are also introduced in each case to the life of the researcher attempting to deal with research participants in honest and just ways.

The middle set of chapters deals with innovative approaches to established methodologies, raising issues of re-examination of productive approaches of issue-concept map, and the creation of a viable sample of informants in online settings. The chapter on researching financial literacy provides an interesting case of researchers grappling with the need to engage with impactful work and what this tells us about the very real and politically significant challenges of researching in a neo-liberal university setting.

The final set of chapters pursues a theme opened in Volume 3 of this series, that of research in education that focuses on the challenges of the Anthropocene, enabling post-human theoretical framings and post-qualitative methodologies. Here we see examples of collaborative, auto-ethnographic research, arts-based responses to the environment, and the possibilities that emerge from challenging traditional research practices using provocations.

As with previous volumes, Volume 4 in the series contains a rich and varied range of methodological offerings that provides insight not only into the challenges of particular methodological approaches as they are used in context but also into the lives and commitments of researchers dedicated to research that makes a difference – to teachers and students in classrooms, to policy, and to our world and shared future.

Peta. J. White
Russell Tytler
Joseph Paul Ferguson
John Cripps Clark

CHAPTER 1

WHAT IS THE PROBLEM WITH STEM EDUCATION? REALISATION OF CRITICAL POLICY ANALYSIS USING AN ARTS-BASED RESEARCH APPROACH

AMANDA PETERS

Abstract

Science, technology, engineering, and mathematics (STEM) is crucial in solving global challenges. Government policies highlight STEM education as key to ensuring a well-prepared future workforce with schools framed as pivotal to this imperative. However, the secondary school curriculum structure of siloed areas remains entrenched whilst student participation in STEM continues to decline. The ongoing challenge of STEM education is contested, with no clear vision for STEM education by policy makers and stakeholders. To understand the complexities of STEM education, a multi-faceted methodology was employed to critically analyse STEM policy, in the Australian context, using a unique Foucauldian inspired post-structural analysis, 'What is the problem represented to be?' (WPR) approach (Bacchi, 2009). The approach highlights how policy represents the problems intended to be addressed, invoking provocations informed by the WPR approach. These provocations were presented to a diverse range of STEM education stakeholders, including students, leading teachers, academics, government representatives and other interest groups, to explore their lived experiences through novel arts-based approaches including drama-like activities. Arts-based approaches provide opportunities to uncover what is silenced, to reveal taken for granted assumptions and gain insight into the complexities of STEM education, complementing the WPR approach. The findings aim to inform future STEM policy and STEM educational reform.

Key words: STEM education, Critical policy analysis, WPR approach, Arts-based approach

STEM and the Australian Education System–the Victorian local context

In Australia, the demand for STEM skills is increasingly important for the national economy. Australians who have studied science, technology, engineering and mathematics (STEM) are positioned as helping to solve a range of future problems across diverse sectors (Australian Government, 2020). The STEM policy agenda in Australia spans all education sectors and systems, including science, innovation, and industry (Freeman et al., 2015). In Australia, STEM refers collectively to a broad field of distinct and complementary approaches to knowledge (Australian Government, 2020). Globally, government policies position STEM as a driver to nation prosperity. Yet, there is no international agreement on the definition of STEM (Fang & Hsu, 2019), with varying definitions depending on the ideas and focus of the STEM goal within the policy.

The importance of building a highly skilled STEM workforce is highlighted within Australian education policies. School education, characterised by STEM, is seen as pivotal in developing students' skills and capabilities to secure future jobs (Jordan, 2018). The Australian National STEM Education Strategy (Education Council, 2015) refers to STEM as the teaching of each of the disciplines, with a cross-disciplinary approach to engage students in STEM fields of work and to improve problem solving and critical analysis skills of students. A 'STEM' definition is positioned as flexible in response to the local context (Australian Government, 2019). STEM adaptability is positively framed (Education Council, 2015), it can provide important connections between core subject areas, it can be its own group of related content areas or it may be considered a meta-discipline providing added value to content areas, encompassing a continuum of concepts (Timms et al., 2018). All offer optimistic interpretations rather than revealing the fragmented and contradictory STEM policies existing between federal, state, and territory jurisdictions throughout Australia (Freeman et al., 2015).

The focus of the research is STEM education policy within the Australian state of Victoria. STEM is a key part of the Victorian curriculum driven by compelling needs at all levels of the education system (Department of

Education and Training, 2016). The aim is to improve STEM education opportunities and outcomes with a strong focus on STEM skills and capabilities, for economic and societal benefits (Department of Education and Training, 2016), as informed by the Australian National STEM School Education Strategy 2016-2026 (Education Council, 2015). The Department of Education and Training (2016) aims to equip teachers and schools to deliver the priority areas of learning, including STEM. To support schools to deliver the priority areas, the government provides a detailed whole school guide to curriculum planning, specifically focusing on content-driven instruction (Department of Education and Training, 2020). However, there is no mention of STEM in the document. Further hampering the delivery of STEM in schools is the structure of the Victorian curriculum, it remains in discrete, discipline-based subjects. Conversely, educators are expected to navigate how to deliver STEM learning in their context with a cross-disciplinary approach. Teachers note the continuous struggle between teaching in authentic contexts, developing the skills necessary for student agility in the future workforce and measuring learning through state-level, discipline based standardised assessments (Bartholomew et al., 2020).

STEM may be viewed as an innovation requiring a shift to well established culture. The government is reticent in realising the struggle for educators to design and implement change (Fullan, 2007) in a traditional, structured, and highly accountable system. Curriculum redesign is notoriously slow with today's curriculum ill-equipped to prepare students for the diverse needs of the future (Organisation of Economic Co-operation and Development (OECD), 2020). In the curriculum, STEM skills and capabilities are, at best, implicit and ill-defined (Siegel & Giamellaro, 2020). Without provision of a STEM curricular approach, the act of authentically teaching STEM in schools, as it is enacted in vocational and professional settings, continues to be out of reach (Blackley & Howell, 2015). The indirect nature of STEM education requires educators, within their local context, to explicitly define, adopt, and deliver STEM in their school (Siegel & Giamellaro, 2020), embedding strategies that encompass the school's routines, structures, and values (Peters-Burton et al., 2014). It would be a brave school in such a climate of accountability and comparison to step away from the curriculum silos and government measure of school performance to trial integrated STEM education. There are many challenges that need to be overcome for integrated STEM education to succeed (Blackley & Howell, 2015).

Interrogating Victorian STEM policy–why the WPR approach?

Australian government policies outline the importance of STEM education in developing a well prepared future workforce, highly skilled in STEM (Jordan, 2018). Policies are socially constructed and are a product of multiple authors and agendas. They are formed with specific cultural, historical, national, and jurisdictional contexts (Ball, 2015). Policy is seen as a solution to STEM education, where policy initiatives and re-thinking the STEM curriculum are deemed necessary to support schools to make an impact on STEM education and improve student outcomes (Timms et al., 2018). However, STEM education is a complex, dynamic process. It is active, co-constructed in highly nuanced, localised contexts (Siegel & Giamellaro, 2020) with no consensus for how STEM education is delivered in the school context within curriculum constraints (Murphy et al., 2019). It is the instability of STEM education and the rejection of the notion it is a fixed entity that aligns itself to critical analysis using the 'What is the problem represented to be?' (WPR) approach (Bacchi, 2009).

The WPR approach to critical policy analysis is a post-structuralist analysis. Rather than taking STEM education for granted as a problem to be fixed by government policy, the analysis shifts to how the problem came to be. The WPR approach seeks to understand how governing takes place by looking at how the proposed solutions problematise an issue, in this case, the issue of STEM education. The premise contends that policies, value laden and contested, embed 'problematisations' (Bacchi, 2009) and governance is through problematisations rather than through policy. The 'problems' are created within the policy-making process, shaping implicit 'problems' rather than addressing them (Bacchi, 2009). The focus of the critical analysis shifts from the proposed solution to scrutinising the issue that has been problematised (refer to Figure 1.1). This opens a different type of inquiry into governance compared to conventional policy analysis (Bacchi, 2009).

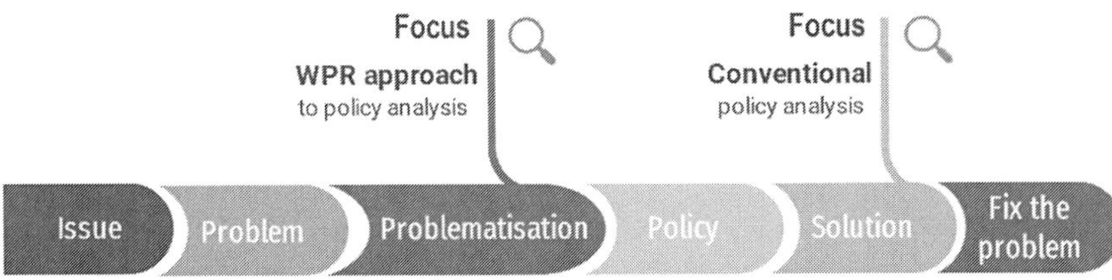

Figure 1.1: Policy analysis–WPR approach compared to conventional policy analysis.

The WPR approach views policy as a discourse, where both solutions and problems are created (Bacchi, 2009). It provides a systematic methodology to interrogate the broader social processes in propagating the problem. In-built assumptions and biases embedded in policy are disrupted, testing truth claims and imagining alternative ways of developing policy and practice (Goodwin, 2012). The WPR approach (Bacchi, 2009) is considered valuable in analysing Victorian STEM education policy, the complexity of STEM education and the impacts on stakeholders. It is underutilised in the field of education (Tawell & McCluskey, 2021) and to date, has not been applied to STEM policy or STEM education. It is contemporary and political, potentially leading to transformation (Goodwin, 2012) within the field of STEM education. The aim is to destabilise the authority of 'problem solving', creating a right to the problem (Deleuze, 1994), to consider how problems are thought about, rather than accepting them.

How did the STEM education 'problem' arise as an issue for governments to address? The WPR approach is key to this research as STEM education is ill-defined with diverse constructions of the STEM 'problem'. The government's narrow understanding of the STEM 'problem' along with their simple solutions (Jordan, 2018) takes precedence as they govern schools and highlight STEM policy as a priority. The STEM 'problems' exist and are lived (Bacchi, 2009) in the system. The findings from the policy analysis will be used to understand the STEM education 'problem'–how it exists and how it is lived by a range of stakeholders in the education system.

Critically analysing Victorian STEM policy using the WPR approach

The WPR approach is used as both methodology and method to facilitate critical analysis of Victorian STEM policy. In education, problem solving is almost hegemonic, with evidence-based policy and the push to produce students as 'problem solvers'. The WPR approach interrupts the presumption that 'problems' are ripe for fixing and highlights 'problem' questioning as a form of critical practice. Applying a specific framework, utilising six questions (Bacchi, 2009) assists in uncovering the complexity of STEM education (Bacchi, 2009). The first question considers what the government is proposing to change to elicit what the 'problem' is. The following questions examine the proposal for change including the rationale and underlying assumptions driving the change. The fourth question identifies the silences in what needs to change and considers the effects of the proposed change (Bacchi, 2009). Six steps are used to critically scrutinise Victorian STEM policy. Each step is explored below.

1. What's the 'problem' represented to be within STEM policy?

The first question is for clarification. It asks to work backwards from the solution, as identifying the solution reveals what needs to change. In turn, this uncovers what the problem is. The question underpins the methodology, and its purpose is to dig deeper than face value of the solution. The current Victorian STEM Education State Plan (Department of Education and Training, 2016) aims to equip "all Victorian learners with STEM capabilities", to "build foundational STEM skills in young children, lift STEM achievement in schools, and ensure our higher education and training sectors are creating a STEM-skilled workforce" (p. 3). The solutions to improve STEM education and skills, as outlined in the STEM Education State Plan (Department of Education and Training, 2016) are summarised below (refer to Table 1.1).

Table 1.1: Problem representations of initiatives in the current STEM Education State Plan (2016)

Initiative (Solution)	What needs to change	'Problem'
Tech Schools built for young people to "discover and develop the skills they need for lifelong learning and future careers" (p.12).	Tech schools are needed to develop students' skills.	Schools (specifically teachers) require support to develop students' skills for the future.
Professional learning programs to "build teacher's capacity to teach and promote STEM" and train teachers to "become experts in STEM teaching and inspire fellow teachers" (p.12).	Train teachers to teach and promote STEM.	Teachers are unable to teach or promote STEM.
STEM facilities, science and mathematics specialist centres, and selective-entry schools "that benefit teachers and students across Victoria" and "inspire a passion for STEM in more students and teachers" (p.13).	Attending specialist centres will inspire teachers and students to be passionate for STEM. Science and select-entry schools are required to benefit some teachers and students.	Most teachers and students are not passionate about STEM. Most schools are not equipped to teach STEM. STEM is not for everyone.

Using the WPR approach to identify the solutions, in this case, initiatives, identifying what needs to change and therefore realising the presumed problems highlight a recurring theme in the policy. A range of initiatives are included to support schools to enhance teaching of STEM education, assuming improved student outcomes will follow. In some cases, schools are bypassed altogether, with the building of new structures, specifically Tech Schools and select-entry schools, aimed at benefiting teachers and students. The underlying problem appears to be the presumption of the current inadequacies of schools, particularly teachers, to effectively teach STEM and develop appropriate STEM education to improve student outcomes.

Reversing the problematisation (Shapiro, 2012) reframes the problematisation back on the Victorian government, positing that the government fails to recognise what is currently occurring in schools. Rather, the initiatives provide tangible solutions of building new structures and providing resources, a neo-liberalist approach to solving a problem. By shifting the problematisation to the issue of governance, the failure of the initiatives to address the challenges of the issues in schools, including accountability and curriculum structure, opens up the possibility of critical examination of schools and teachers as governable subjects. The policy making process highlights the presumption that schools, and teachers are unable to teach STEM and need to be 'fixed'.

2. *What government assumptions underlie this representation of the 'problem'?*

The second question explores unexamined ways of thinking and uncovers the deep-seated cultural values underpinning policy, requiring a specific understanding of social relations. In STEM education, this refers to the cultural and ideological influence of how schools and teachers are generally perceived in Australian society, within the Victorian context, and how they are connected to the larger socio-cultural context. Commonly accepted authoritative knowledges that determine what is true in our society are interrogated (Bacchi, 2012). The Victorian STEM Education policy (Department of Education and Training, 2016) assumes schools and teachers are unable to teach STEM and require support. The underlying assumption is that learners don't possess the required STEM skills for the future workforce. Most of the solutions are targeted at providing teachers with a range of pathways and opportunities to learn how to teach STEM. STEM centres are built to develop students' skills, separated from the local school context. These centres are intended to inspire a passion for STEM for teachers and students. It follows that students need outside sources, separate from teachers, to develop STEM skills, as teachers are unable to do this. It presumes STEM is required to be taught in a range of specialist centres and select-entry schools, seemingly to safeguard against the inadequacy of STEM education in schools. Rather than teachers and schools being the 'problem', it is argued the 'problem' is the government's lack of focus on the institutional processes and accountability within the education system.

The policy presumes teachers have control over what they teach, focusing on teaching in response to the needs of the future STEM workforce. It assumes teachers understand this and will embrace the opportunity to build their STEM teaching capacity, 'fixing' the problem of teachers not being able to teach STEM and 'fixing' the problem of students' underachievement in STEM. The policy fails to refer to the issues the political push of STEM into education uncovers. It neglects to acknowledge the prescribed discipline-based curriculum structure in Australian schools (Freeman et al., 2019) with barriers, institutionally and systematically, to implementing STEM education (Tytler et al., 2019). The policy ignores the complexities of delivering STEM education. It highlights it is not a one size fits all approach.

3. How has this representation of the 'problem' come about?

The third question comprises a Foucauldian approach where power dynamics in cultural and/or historical contexts are considered. In the context of critical policy analysis, it seeks to uncover the actors who have shaped the policy making (Bacchi, 2012). In Australia, STEM skills are positioned as the saviour, solving a myriad of future problems impacting on the national economy. Subsequently, the school sector is encumbered with the economic aims of government and industry (Barkatsas et al., 2018). The Victorian Government (2016) is "committed to delivering jobs and prosperity to Victorians and has a strong plan to grow the economy. The key to a prosperous future lies in a highly skilled workforce, including strong capability in STEM" (p. 3). Governments fund education, where the education system is "set to play a critical role in meeting tomorrow's employment challenges. By transforming how and what we teach our children and young people" (p. 20). However, the education system is mandated to teach a Victorian curriculum with no mention of STEM in the curriculum planning and reporting guidelines (Department of Education and Training, 2020).

The Victorian Curriculum and Assessment Authority (VCAA) considers STEM an integrated curriculum approach yet acknowledges the segregated teaching structures that secondary schools employ is seemingly in opposition to teaching STEM (Victorian Curriculum and Assessment Authority, 2021). In the policy, teachers are presumed to be the problem in not being 'transformative' in teaching and responding to the needs of the skill set of future workforces. The policy offers many solutions to 'fix' the problem of teachers. However, is the problem the teachers or is it the system?

4. *What is left unproblematic? Where are the silences? Can the 'problem' be conceptualised differently?*

The questions aim to uncover what isn't said in the discourse, to think about the problem differently. Attending to the silences in the problematisation refocuses attention on how and why some subjects are problematised while others are not (Pringle, 2019). The STEM Education State Plan (Department of Education and Training, 2016) claims that teachers and schools will be equipped "to deliver the priority areas of learning, including STEM" (p. 13) and "implementation of the Victorian Curriculum which supports improved achievement in STEM…for Victorian schools, STEM is a fundamental part of the Victorian Curriculum" (p. 3). Yet there is no mention of STEM in the Victorian Curriculum or the historical structure of secondary schools teaching subjects in isolation. "Lifting STEM proficiency in students requires developing STEM proficiency in our teaching workforce. This includes ensuring an adequate supply of teachers with contemporary content knowledge in STEM subjects, and the pedagogical skills to stimulate their students' interest and learning" (p. 5). However, discipline-based content knowledge is the focus of curriculum, yet skills are espoused to lead to economic prosperity. Teachers are assumed to be the problem and students the result of that problem, yet both are silenced in the policy. There are also silences in the disregard for the historical and cultural contexts of secondary schools including curriculum, school structure, and teaching qualifications in discrete learning areas. The policy rhetoric is seemingly disconnected with the reality of teaching and learning in schools.

5. *What effects are produced by this representation of the 'problem'?*

The fifth question addresses representation, meaning and effect (Pringle, 2019). The importance of STEM is paramount in schools, yet STEM education implementation varies significantly. How schools and teachers perceive and interpret the problem, and how the problem impacts on themselves and the students are considered. Who benefits from the problem? Who might benefit if the problem is differently represented? The approach is relevant to the importance of critical research considering the material and lived effects that discourse, policy, or governance may have on the subjects within the policy (Bacchi, 2012). In regard to the discursive effects and the limitations of what is said and placed on the problem, the Victorian STEM education policy (Department of Education and Training, 2016) focuses on the need to equip learners with STEM skills for the future workforce for the prosperity of the state. STEM is considered a priority, privileging STEM subjects and skills over non-STEM subject areas, creating disparities. The policy assumes teachers regard STEM as a priority, and require support to teach STEM. Even with support, it is assumed students will require access to special built STEM centres to learn STEM skills. The myriad of solutions inadequately addresses the problem the policy creates–the assumption teachers want to teach STEM, are unable to teach STEM, so STEM teaching is required to be outsourced. Rather, the implications realise the significant de-valuing of teachers in policy, a snapshot of the wider societal assumption.

How might teachers perceive and interpret the problem? How does this shape an understanding of themselves and the issues? With respect to the policy, teachers (subjects) are offered numerous ways to develop their capacity to be STEM teachers. Additionally, the purpose-built STEM facilities to develop students' skills, without their teachers teaching them; the promise of increasing the proportion of STEM graduates as teachers, the integration of outside organisations to invigorate STEM, and the lack of achievement of students can only be detrimental to current teachers. The current beneficiaries of the 'problem' of STEM education are government and employers as they focus on building the economy and the future workforce. If the problem is newly represented, educators and students may also benefit.

6. ***How/where has this representation of the 'problem' been produced? How has it been (or could it be) questioned, disrupted, and replaced?***

The sixth question considers the origins, purposes, and effects of the policy. This question is informed by the third question, how has the problem come about to de-stabilise taken for granted assumptions (Bacchi, 2012). Within the policy there are a range of assumptions including, the purpose of education to equip learners with STEM skills for employability, teachers as the problem in declining student achievement in STEM, and STEM proficiency best measured using national standardised testing for numeracy and international testing for mathematical literacy. The assumptions provide impetus for provocations, including: Is equipping learners with STEM skills the answer? What is the purpose of education? What barriers are schools, teachers, and students facing in STEM education? How might the system be re-considered? How else might STEM proficiency be determined?

The WPR approach provides a critical, consecutive breakdown of social, economic, political, cultural, and historical factors influencing and shaping Victorian STEM education policy. It focuses away from economic drivers influencing education and spotlights the marginalised stakeholders, specifically teachers and students. Proposed solutions target improving teachers and students (the 'problem'). Assumptions of de-valued teachers and the historically and culturally inert education system are uncovered. The WPR approach provides the impetus to move beyond interpretation and towards challenging the status quo. However, the WPR approach refrains from addressing how policy problems are realised in practice (Clarke, 2019). In this research, the WPR is supplemented with an arts-based research (ABR) approach to explore the lived experiences of policy in practice. Are the effects of problematisations uncovered in the critical analysis realised in practice? Are they experienced by STEM education stakeholders? If so, how are they experienced?

An arts-based research approach to explore the realisation of STEM policy in practice

The WPR approach infers the effects of policy from policy analysis. To understand the lived impacts of policy through first-hand experiences of stakeholders, the ABR approach is employed as it has synergies to the WPR approach. It seeks to deconstruct complexity, uncover assumptions, enabling a voice to all participants, to rethink what is possible. To gain further insight into the impact of the problematisations in practice, provocations from the STEM education policy analysis were used to develop drama-like activities, informed by the ABR approach. Drama-like activities afford creativity, infinite possibilities, and a lasting learning experience. A diverse range of STEM education stakeholders, approximately fifteen, were invited to a focus group workshop to participate in drama-like activities, co-facilitated by two researchers. The stakeholders included students, leading teachers, academics, and government representatives. The activities provided the opportunity for the participants to reflect on the multiple possible meanings (Raphael & O'Mara, 2002) and experiences of STEM education.

An arts-based approach is enacted to complement the WPR analysis. Conventional research relies on discourse to verify claims and reduce ambiguity. Methods such as individual interviews rely on words, limiting what is expressed or represented, neglecting what is omitted. Quantification of data leads to propositional claims and a reductionist approach to research, unhelpful in this case, where the complexities of STEM education are intended to be uncovered. Instead, arts-based research broadens our understanding of the ways in which people come to know. It supports interpretive studies and extends freedom, democracy, and political voice into everyday lives (Finley, 2011), aligning with the premise of the WPR approach. Methodological pluralism is an important virtue to garner multiple perspectives on how one sees and understands the world (Barone & Eisner, 2011). Arts-based research enables complex and nuanced interactions and differences between people to be noticed in various forms of representation. It is important to notice these interactions and differences to understand the perspectives of stakeholders in STEM education. These differences are recognised, leading to challenges, enabling debate and discernment. Arts-based research provides the opportunity to deepen understanding and expose the silences, supporting the WPR approach of critical policy analysis.

Arts-based approaches assist in uncovering what is hidden, to engage and recognise the inner being and knowing of people. In educational research, there is an imperative to innovate methodology, to broaden the tools used to represent the world and to expand what defines the meaning of research (Barone & Eisner, 2011). In my work as a teacher educator in the tertiary sector, I have employed arts-based approaches in teaching curriculum study, supported by researchers in this field. I have come to understand that arts-based research is a productive heuristic that enlightens, seeking credibility rather than validity (Barone & Eisner, 2011). The practice creates liminal openings where people can "imagine new visions of dignity, care, democracy, and other decolonising ways of being in the world. Once it has been imagined, it can be acted upon, or performed" (Finley, 2011, p. 443). In this research, it provides an avenue to delve into and enhance the WPR approach, particularly the opportunity to understand the effects and 'disrupt' the 'problem' of STEM education.

An arts-based approach is an evocative and useful frame for this research as it promotes further inquiry, to make conversation more interesting and accessible with a range of stakeholders. It enables empathic participation to explore meanings of STEM education, particularly participants' affective experiences (Finley, 2011). The approach exploits expressive form to capture what people know and how they understand or see the world. Metaphors and analogies facilitate the opportunity to share experiences and express ideas and meaning. The approach emphasises the importance of feeling to gain a greater understanding of people and their situations. It goes beyond quantitative measure to pursue an expressive form to deepen human understanding of the STEM education issue (Barone & Eisner, 2011).

Both the arts and sciences are occupied with the processes of questioning and knowing, to open to new ideas. They convey diverse images enabling us to notice what we haven't previously seen. The arts teach us to see, feel, and know. An arts-based approach provides expression and utility in research. It exposes differences, or vexing, in knowing and how we come to know (Barone & Eisner, 2011). The vexing, particularly grounded in the issue of STEM, where there are unresolved tensions, serve to animate and explore. It promotes scientific inquiry, making the unnoticed noticeable, expanding awareness. It enables this research to extend beyond discursive communication and express meanings in a diverse range of other forms. New perspectives are provided, exploring an approach to representation grounded in the aesthetic, supporting the STEM policy analysis to conceptualise the 'problem' differently. It proposes to deepen understanding of STEM education, creating extensive insight.

Different forms of representation afford an awareness of human understanding that the spoken word, for example in interviews, cannot fully capture. Representations may be discursive and nondiscursive. They are non-redundant, making different forms of understanding possible (Barone & Eisner, 2011). Employing this approach enables the complexities of STEM education to be viewed in many ways. Arts-based research seeks to capture the plurality of these views. It moves away from the notion of reaching a final understanding or consensus, the conventional assumption of research. Instead, it encourages stakeholders to see the world from a different perspective and calls into question a singular knowing or orthodox point of view. Ambiguities are presented, encouraged to peer beneath the familiar, the obvious, to take a risk and re-scrutinise what is known and comfortable.

ABR is highly useful in seeking understanding, and potential emancipation of participants, aligning with the premise of the WPR approach. Taken-for-granted premises and ways of knowing in STEM policy are questioned with a view to moving in different directions, to look at and experience the world in a new way. It is through experiencing, rather than discourse or argument, that the acceptance of alternative meanings of issues and practices may be understood. The use of expressive design elements and a variety of forms of representation enriches what one knows, to uncover questions buried within proposed answers, in STEM policy. The opportunity to re-experience aspects of the world previously unnoticed is provided. Encouraged to free ourselves of entrenched ways of knowing how we experience STEM, participants are asked to reflect, re-consider, or re-make their understanding of the issue, to generate profound questions rather than correct answers (Barone & Eisner, 2011). ABR approaches are certainly well suited to supplement the WPR approach.

Exploring the ABR informed activities

ABR is used to think more critically about new forms of becoming in the world, to think and act differently. ABR complements WPR as it provides the opportunity for discourse, creativity, freedom, and political voice (Finley, 2011), to disrupt master narratives that maintain the status quo. The focus group provided the opportunity for interaction and an element of synergy between the participants (Denscombe, 2014), not afforded by individual interviews. The aim was not to reach consensus, rather to uncover a range of experiences and perspectives (Hennink, 2014), through engaging in drama-based activities in the liminal space of the drama studio. Purposeful drama activities were selected, informed by the WPR analysis, to provide a participant centred, social-constructivist approach to gain insight into the issues of STEM education (Dorion, 2009). Episodic approaches enabled embodiment of the moment, resisting a narrative form, enabling relevant activities to be enacted at appropriate stages (Raphael & O'Mara, 2002).

It can be challenging for participants to be in an unfamiliar space. The warm up activities seek to enable a permissive, non-threatening environment, for participants to feel more comfortable to share their views (Hennink, 2014). The warm up activities were designed to be non-verbal, to be emergent and inclusive, using the body to explore and enact meaning and perspective, realising there is not one truth (Finley, 2011). Ten second constructions provided the opportunity for groups to embody and represent STEM education and the education system (refer to Figure 1.2), each in ten seconds. Participants explained their reasoning, to uncover what is explicit, hidden or silenced in STEM education, to make the familiar strange and recognise multiple meanings. The education system was embodied by participants, demonstrating their perspectives of the inert institutional structure, the all-encompassing governance, the diverse range of stakeholders, the disconnect, the power imbalance and the silenced. A snapshot encapsulating stakeholder perspectives, highlighting comparable findings from the WPR approach to STEM education policy analysis.

Figure 1.2: The Education System

The following activity, 'Role on the Wall', is a powerful tool, well suited to intergenerational contexts (refer to Figure 1.3). It provided a safe context in which to uncover assumptions and explore the otherness of participants (Houseal et al., 2013), well suited to complement the policy analysis. It provides an innovative framework for responding to questions within the WPR approach. Participants were encouraged to write viewpoints, thoughts, and feelings on blank silhouettes on sheets of paper, highlighting stakeholder roles. They had time and space to consider their own perspectives not only as a stakeholder, but also their perspectives on other stakeholders, in conjunction with reading other contributions. The activity enabled representation of a composite viewpoint, compressing group reflection on a stakeholder into a visual record. Debriefing afterwards provided the opportunity to voice noticings and wonderings of interest. To hold the space and not aim to solve the problem, to make thinking clearer, to render the richness and complexity of their experiences and perspectives (Cahnmann-Taylor & Siegesmund, 2017).

The prompts, informed by the WPR approach to STEM policy analysis include:

- Who makes, implements, and monitors STEM policy?
- How do the subjects of STEM policy interpret, enact, and experience the policy in the school and/or classroom?
- What are the problems of the policy?
- What isn't said in the policy?

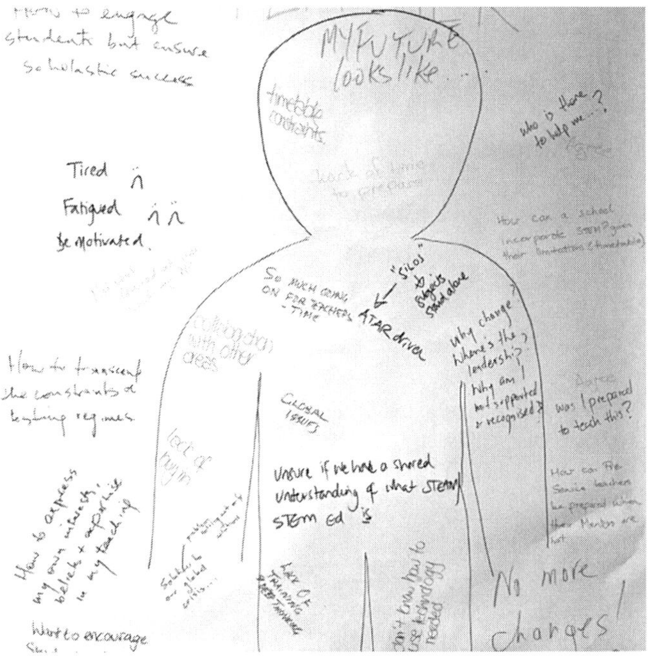

Figure 1.3: Role on the Wall, stakeholder perspectives of STEM

The metaphor activity asks participants to create (using plasticine) a metaphor for their experience of STEM education and their vision for the future of STEM education. The use of metaphor is used to reflect what is not obvious, to deepen understanding of what is abstract (Breslin & Camacho, 2021), to uncover the complexities of STEM education. Metaphor enables understanding of how reality is constructed, recognising new perspectives. It allows the imagination to work at an unconscious level and surfaces different understandings, carrying a message beyond formal explanation, encompassing multiple layers and providing the opportunity for people to connect with the experience in different ways (Simons & McCormack, 2007).

Participants created their sculptures and engaged in a gallery walk. They were asked to consider and discuss others' creations followed by each artist sharing their meaning, promoting discussion to understand other perspectives and socially construct meaning. Visual images are a powerful communicative tool with the ability to assist in seeing things in different ways. The aim was to discover, explore, and illuminate the realisation of policy in practice. I wonder how you might interpret the sculpture of STEM education (refer to Figure 1.4)?

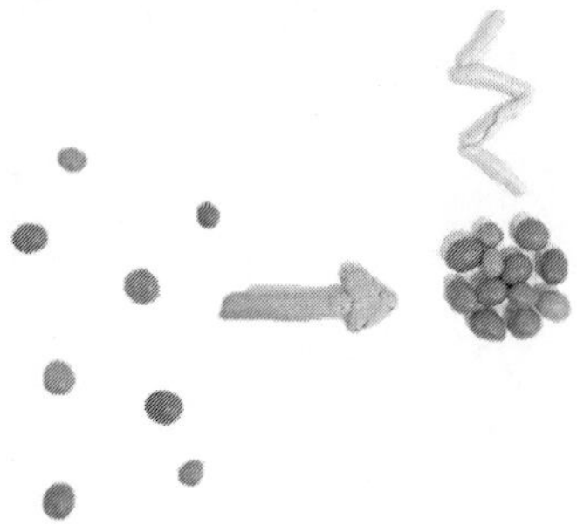

Figure 1.4: Sculpture representing the experience of STEM education, currently and into the future

The artist's meaning included the current siloed approach to STEM education in schools, with mostly boys doing STEM. Into the future, STEM education, for students, would be more relevant to real life, incorporating skill sets across disciplines. It would be gender inclusive, with everyone working together to hear different perspectives. The meaning provides insight into two questions posed by the WPR approach, namely how can the 'problem' be conceptualised differently and how it could be disrupted and replaced.

The links to the WPR approach of STEM policy analysis include:

- Whose voice is missing in STEM education? What isn't said about STEM education in schools?
- How is STEM education experienced?
- How could STEM education be experienced and/or re-considered?

Consider further sculptures representing experiences of current STEM education and into the future (refer to Figures 1.5-6). How might you interpret these sculptures?

Figure 1.5: A secondary student's representation of STEM education experiences, currently and into the future

Figure 1.6: A teacher's representation of STEM education experience, currently and into the future

The artists' meanings of the representations of STEM education now and into the future included each subject currently taught separately, with the feeling of being overwhelmed, moving towards intertwining of disciplines, supporting each other. All disciplines coming together, a sense of connection and community, that can be bright and brilliant.

Conclusion

The WPR approach provides a novel way of thinking, to shift the mindset of policy analysis presuming 'problems' can be easily fixed with solutions, to a critical interrogation of problem solving knowledge (Bacchi, 2009). In Victorian STEM education policy, the solutions are focused on improving teachers and students, those silenced within policy. Assumptions of de-valued teachers and the historically and culturally inert education system are uncovered. The WPR approach provides the impetus to move beyond interpretation and challenge the status quo. The effects of discourse on the lived impacts of policy are acknowledged, and how the 'problem' evokes stakeholders' perceptions of themselves are implied. Although, how the policy 'problems' actually play out in practice is a limitation within the WPR approach (Clarke, 2019).

Quantitative measurement is considered essential to improve education (Stéphan et al., 2019) yet it is hardly adequate to capture the complexity. The research proposes supplementing the WPR approach using ABR approaches, due to their complementary nature. They both have an inherent emancipatory function, to uncover the 'others' and disrupt taken for granted assumptions. The ABR approach provides a natural heuristic to delve deeper into the WPR analysis findings, particularly how they are experienced by a range of stakeholders. Learning in drama moves beyond the individual knowledge and experience, to consider broader contexts and wider perspectives, providing the impetus for new insights (Raphael & White, 2021), to understand how the 'problem' with STEM is represented. Moreover, ABR enables greater subjective approaches to analysis and interpretation. Drawing on an interpretivist approach, WPR guides the thematic analysis, to articulate the entanglements and complexity within STEM education. As Barone and Eisner (2011) state, the end in view is not to arrive at a singular and unchallengeable slice of knowledge; it is to generate questions to make conversations more interesting. The WPR approach, ably supported by the ABR, further disrupts the current rhetoric, situating the STEM education debate into the wider debate on the purpose of education in our society.

Acknowledgements

I would like to acknowledge and thank my supervisors, Associate Professor Peta White and Dr. Jo Raphael for their insights, support, patience, guidance, and expertise in developing this chapter.

References

Australian Government. (2019). *STEM school education interventions: Synthesis report.*

Australian Government. (2020). *Australia's STEM workforce*. Canberra.

Bacchi, C. (2009). *Analysing policy: What's the problem represented to be?* Pearson.

Bacchi, C. (2012). Why study problematizations? Making politics visible. *Open Journal of Political Science*, *2*(01), 1.

Ball, S. J. (2015). Education, governance and the tyranny of numbers. *Journal of Education Policy*, *30*(3), 299-301.

Barkatsas, A., Carr, N., & Cooper, G. (2018). *STEM education: An emerging field of inquiry*. Brill Sense.

Barone, T., & Eisner, E. (2011). *Arts Based Research*. SAGE Publications.

Bartholomew, S. R., Strimel, G. J., Lucietto, A. M., & Akdere, M. (2020). Reinventing high school: Understanding the challenges and successes of transforming education to meet student, society, and industry needs. *Journal of Technology Studies*, *46*(1).

Blackley, S., & Howell, J. (2015). A STEM narrative: 15 years in the making. *Australian Journal of Teacher Education*, *40*(7), 102-112.

Breslin, S., & Camacho, M. M. (2021). Metaphors of change: Navigating a revolution in engineering education. *Engineering Studies*, *13*(1), 53-77.

Cahnmann-Taylor, M., & Siegesmund, R. (2017). *Arts-based research in education: Foundations for practice*. Taylor & Francis Group.

Clarke, A. (2019). Analyzing problematization as a situated practice in critical policy studies: A case study of 'customer focus' policy in urban compliance services. *Critical Policy Studies*, *13*(2), 190-210.

Deleuze, G. (1994). *Difference and repetition*. Columbia University Press.

Denscombe, M. (2014). *The good research guide: For small-scale research projects* (5th Ed.). Open University Press.

Department of Education and Training. (2016). *STEM in the Education State*. Melbourne: Victoria State Government.

Department of Education and Training. (2020). *A whole-school guide to curriculum planning*. Melbourne: Victoria State Government.

Dorion, K. R. (2009). Science through drama: A multiple case exploration of the characteristics of drama activities used in secondary science lessons. *International Journal of Science Education*, *31*(16), 2247-2270.

Education Council. (2015). *National STEM school education strategy 2016-2026*. Canberra: Australian Government

Fang, S.-C., & Hsu, Y.-S. (2019). Assessment challenges in STEM reforms and innovations. In Y.-S. Hsu & Y.-F. Yeh (Eds.), *Asia-Pacific STEM*

teaching practices: From theoretical frameworks to practices (pp. 191-203). Springer Singapore.
Finley, S. (2011). Critical arts-based inquiry: The pedagogy and performance of a radical ethical aesthetic. In N. Denzin & Y. Lincoln (Eds.), *Sage handbook of qualitative research* (4th Ed., pp. 435-450). Sage.
Freeman, B., Marginson, S., & Tytler, R. (2015). *The age of STEM: Educational policy and practice across the world in science, technology, engineering and mathematics*. Routledge.
Freeman, B., Marginson, S., & Tytler, R. (2019). An international view of STEM education. In A. Sahin & M. Mohr-Schroeder (Eds.), *STEM Education 2.0: Myths and truths - What has K-12 STEM education research taught us?* (pp. 350-363). Ringgold.
Fullan, M. (2015). *The new meaning of educational change*. Teachers College Press.
Goodwin, S. (2012). Women, policy and politics: Recasting policy studies. *Engaging with Carol Bacchi: Strategic interventions and exchanges*, 25-36.
Hennink, M. M. (2014). *Focus group discussions*. Oxford University Press.
Houseal, J., Ray, K., & Teitelbaum, S. (2013). Identifying, confronting and disrupting stereotypes: Role on the wall in an intergenerational LGBTQ applied theatre project. *Research in Drama Education, 18*(2), 204-208.
Jordan, K. (2018). Reading STEM as discourse. In *STEM education: An emerging field of inquiry* (pp. 45-60). Brill.
Murphy, S., MacDonald, A., Danaia, L., & Wang, C. (2019). An analysis of Australian STEM education strategies. *Policy Futures in Education, 17*(2), 122-139.
Organisation of Economic Co-operation and Development (OECD). (2020). Curriculum (re)design: A series of thematic reports from the OECD Education 2030 project.
Peters-Burton, E., Lynch, S., Behrend, T., & Means, B. (2014). Inclusive STEM high school design: 10 critical components. *Theory Into Practice, 53*(1), 64-71.
Pringle, W. (2019). Problematizations in assisted dying discourse: Testing the "what's the problem represented to be?" (WPR) method for critical health communication research. *Frontiers in Communication, 4*.
Raphael, J., & O'Mara, J. (2002). A challenge, a threat and a promise: drama as professional development for teacher educators [Paper in: Drama and Learning, Donelan, Kate and Cahill, Helen (Eds.)]. *Melbourne Studies in Education, 43*(2), 77-86.
Raphael, J., & White, P. J. (2021). Transdisciplinarity: Science and drama education developing teachers for the future. In *Science and drama:*

Contemporary and creative approaches to teaching and learning (pp. 145-161). Springer.

Shapiro, M. J. (2012). *Discourse, culture, violence.* (Vol. 3). Routledge.

Siegel, D., & Giamellaro, M. (2020). Defining STEM within a school district: A co-constructed and evolving process. *Cultural Studies of Science Education, 15*(3), 739-773.

Simons, H., & McCormack, B. (2007). Integrating arts-based inquiry in evaluation methodology: Opportunities and challenges. *Qualitative Inquiry, 13*(2), 292-311.

Stéphan, V.-L., Joaquin, U., Soumyajit, K., & Gwénaël, J. (2019). *Measuring innovation in education 2019: What has changed in the classroom?* Éditions OCDE / OECD Publishing.

Tawell, A., & McCluskey, G. (2021). Utilising Bacchi's what's the problem represented to be? (WPR) approach to analyse national school exclusion policy in England and Scotland: A worked example. *International Journal of Research & Method in Education*, 1-13.

Timms, M. J., Moyle, K., Weldon, P. R., & Mitchell, P. (2018). *Challenges in STEM learning in Australian schools: Literature and policy review.* Australian Council for Educational Research.

Tytler, R., Bridgstock, R., White, P., Mather, D., McCandless, T., & Grant-Iramu, M. (2019). *100 jobs of the future* (064699509X).

Victorian Curriculum and Assessment Authority. (2021). *STEM frequently asked questions*. Victoria State Government.

Amanda Peters (BSc, GDipEd (SecMathSc), MEd) - Deakin University, School of Education, 221 Burwood Highway, Burwood, 3125, Victoria, Australia.

Amanda Peters is a PhD candidate and science education lecturer at Deakin University. She is the Project Manager for the Graduate Certificate of Secondary Science/Mathematics, a course designed to upskill out-of-field science and mathematics teachers to be in-field teachers. Her interests are in STEM education, critical policy analysis, and teacher education. Amanda is a secondary science and mathematics teacher with extensive experience in leading and teaching in secondary schools. She is involved in VCE curriculum and assessment at the state level.
https://orcid.org/0000-0001-9119-7993

CHAPTER 2

SHIFTING ASSESSMENT CULTURES IN SCHOOL SCIENCE

RUSSELL TYTLER, PETA J. WHITE & SHEFALI SHARMA-WALLIS

Abstract

We describe the methodology designed to develop, trial, and validate science inquiry assessment (SIA) tasks in schools with the intent of shifting school culture about assessment and inquiry skills. This research was complicated by the fact that targeted assessment of inquiry is not (often) practised in primary schools, such that the project aims were to include: development of instruments reflecting curriculum descriptors; validation in a practical sense of the possibility of organising and enacting inquiry activities; enticement of teachers to implement engaging inquiry activities; and attention to teachers' perceptions of inquiry skills and commitment to an assessment as distinct from an immersive inquiry program. In response to these multiple contextual constraints the refining and validation of tasks became an iterative process that relied on detailed methodology and careful data management.

Key words: Assessment, Inquiry skills, Teachers and school culture, Primary science education

Introduction

Inquiry practice has a long history as a key feature of science curricula, dating back to the writings of John Dewey (1933, 1996) and Joseph Schwab (1962, 1965). Schwab argued that science curricula should extend beyond the prevailing focus on the knowledge products of science to include the way such knowledge is developed and validated. Jonathan Osborne has more recently pointed out that:

> Four decades after Schwab's (1962) argument that science should be taught as an 'enquiry into enquiry', and almost a century since John Dewey … advocated that classroom learning be a student-centred process of enquiry, we still find ourselves struggling to achieve such practices in the science classroom. (Osborne, 2006, p. 2)

The science education research literature repeatedly characterises mainstream science teaching and learning as consisting of teacher delivery of prepared knowledge, through limiting pedagogies that fail to stimulate the development of higher order thinking (Capps & Crawford, 2013; Duggan & Gott, 2002; Edwards & Mercer, 1987; Goodrum et al., 2011; Goodrum et al., 2001). Even with a long tradition of practical work in science, experimental activities consist largely of explicit instructions that reduce students' experience of inquiry to procedural protocols, with results mainly aimed at illustrating knowledge that is already determined (Holmes & Wieman, 2016). In such activity, students are not often challenged to draw on higher order thinking and decision-making characterising the epistemic practices of science (Duschl, 2008). This has the effect of promoting a misleading view of the relationship between experimental work and scientific knowledge production (Carey et al., 1989).

Writing about inquiry processes in science is bedevilled by the fact that the term is broadly used and represents a range of pedagogies that are generally reformist and student-centred (Anderson, 2002; Chen & Tytler, 2017). However, there are specific forms of inquiry that refer to a set of investigative, epistemic practices of science such as investigative design, measurement, data modelling and analysis that are more clearly defined. These are increasingly represented in curricula internationally, as part of an emphasis on higher order thinking and understanding of the nature of scientific endeavour, for instance in the U.S. through the *Next Generation Science Standards* (National Research Council, 2013), or in Australia through an Inquiry Skills strand (ACARA, 2022).

Within such school systems, there is increasing emphasis on accountability measures for curriculum and teaching and learning, requiring the development of valid, reliable testing regimes. In many ways these demands tend to preclude assessments that are contextual and complex, acting against the trend towards higher order competencies. There is, however, a history of assessment development focused on performance (Palm, 2008). Such measures at system level tend to be expensive in terms of development, administration, and results analysis. However, trends in educational assessment emphasise a more generative role than the traditional summative focus, with assessment 'for' learning (assessment helps teachers plan learning steps), and 'as' learning (where teachers and students jointly monitor learning as integral to the learning design process) coming into increasing prominence as supportive of the development of responsive teaching and learning practices.

> The fundamental purpose of assessment in education is to establish and understand where learners are in an aspect of their learning at the time of assessment. (Masters, 2014, p. 1)

Given this focus on formative assessment, there is a need to develop science inquiry assessments that are situated within school classrooms, to a) assess and support students' development of science inquiry practices, and b) support teachers' capability to recognise and reliably assess such practices. Developing effective assessment regimes is important for teachers' professional learning, to establish a language and pedagogy to support students' development of inquiry practices. In this chapter, we describe the development and validation of a Science Inquiry Assessment instrument for primary schools that is administered and assessed by the classroom teacher.

We discuss methodological issues and approaches to developing such an assessment, particularly focusing on the relative importance of, and tension between the constructs of reliability and validity for assessment. Broadly speaking, these two terms are defined thus (New Zealand Ministry of Education, n.d.):

> The reliability of an assessment tool is the extent to which it consistently and accurately measures learning.
>
> The validity of an assessment tool is the extent by which it measures what it was designed to measure.

We explore, through the research, the methodological implications of developing and trialling a reliable and valid test in a context where the teachers' understandings of the intricacies of scientific inquiry practices are challenged, in recognising these practices and validly judging them.

The Science Inquiry Assessment (SIA) tasks

The original version of the Science Inquiry Assessment (SIA) tasks was developed by the Victorian Department of Education (DE) in 2013 and used as a component of the Primary Mathematics and Science Specialist (PMSS) initiative to build teacher knowledge and skills to assess students' growth in science inquiry. There were two tasks (A and B) for each grade level, each focusing on three of the five Inquiry skills of the Victorian Curriculum: questioning and predicting; planning and conducting; recording and processing; analysing and evaluating; and communicating. The SIA tasks then underwent a series of progressive updates over the intervening years including those related to: the adaptation to the then new Victorian curriculum which occurred in 2019; and then this validation and revision which was contracted late in 2021. The research to perform the validation was hampered by COVID in 2022 and finally the tasks were released to schools in 2023. This latest review involved further refining of the tasks and assessment rubrics, validation of the assessment with classroom research, and the development of student worksheets and completed work samples with clarification of teacher process and scaffolded assessment.

Revisions were proposed based on a desktop review, researcher expertise and experience, and feedback from some of the PMSS Science Specialists, including group discussions about the validity of the task judgment as a basis for diagnostic and then summative assessment. While the tasks were appreciated, there were concerns about the practicality of some tasks in terms of access to equipment, science understandings and inquiry skills of the teachers and students, the importance of the science context in relation to the science inquiry practice, the time required for task completion, and the ability to assess students' inquiry skills via these tasks.

The SIA tasks are designed as classroom tools to:

- situate the inquiry in meaningful contexts that engage students to the point of eliciting their best practice.
- be practical in terms of time, equipment, and teacher knowledge.
- include rubrics that are understandable by teachers making judgments and that discriminate between levels.
- negotiate the tension between teacher support of students to ensure they engage meaningfully at their level, and the need to make judgments about student performance.

The SIA tasks attempt to support teachers to address:

1. A history of little attention paid to inquiry processes and assessment–activities and assessments tend to focus on substantive science concepts; and

2. A history of little explicit education regarding inquiry practices.

In this chapter we explore the following research questions:

1. What are the complexities of generating an assessment instrument and approach that pays dual attention to teacher development and student assessment?
2. How can validity be best thought of in constructing such an instrument and approach?

Research methodology

The project used a design-based research (DBR) methodology (Brown, 1992; Cobb et al., 2003; Reeves, 2006) involving iterative cycles of design of the tasks, trialling in classrooms, then revision/redesign based on analysis of the findings concerning design effectiveness. The DBR cycle is represented in Figure 2.1.

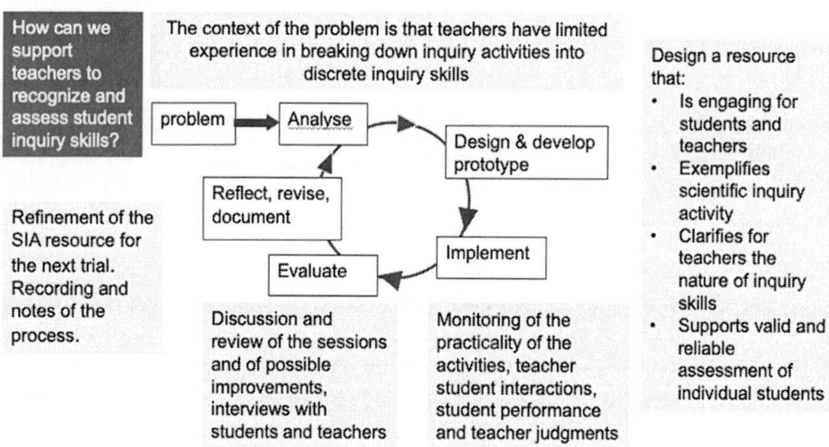

Figure 2.1: The DBR iterative research cycle (based on Stillman et al., 2009 and Plomp, 2013)

The research initially involved the identification of the 'problem' of lack of representation of science inquiry processes in the primary school curriculum and lack of experience of teachers in focusing on and assessing inquiry practices. This general problem identification and prototype design occurred prior to the research team becoming involved, but the new problem that had emerged was the lack of practicality of some of the assessment activities in the length of time involved, lack of precision in the rubric descriptors, and lack of support resources for teachers to interpret the resource. The team thus developed a revised prototype SIA design, and trialled and refined that over two cycles in each of two schools, making specific revisions/refinements after each cycle.

Design-based research has been described as "one of the most exciting evolutions in research methodology of our time" (Christensen & West, 2018, p. 483), being directed to optimising solutions to tangible problems in specific contexts (Reeves, 2006). Cobb et al. (2003) describe 'design experiments' as being a powerful solution in education because of the particularly close connection between theory and practice in recognising and responding to the complexity of teaching and learning. They view DBR as leading to theoretical development and understanding in relation to learning ecologies, in our case the ecology of the inquiry assessment activities involving teacher-student interactions and teacher judgments. The theories developed through design experiments (DBR) are intensely practical, and through the methodology are subjected to on-the-ground reality testing in terms of their pragmatic import for informing practical refinements (Cobb et al., 2003; Tytler et al., 2021).

In the case of the SIA, the theoretical insights we aimed to develop were those relating to validity and reliability in measurement, relating to the identification and clarification of inquiry practices in science classrooms. These developments were tested against the grounded reality of teachers' experience and their capacity to make informed judgments of students' inquiry skills.

Research design

The research design had the following components that were conducted in the order listed:

- Interviews with current and past PMSS teachers who had applied specific tasks. They were invited to share thoughts on the SIA tasks' practicality, the clarity of rubrics, and the possibility of discrimination across the class.
- Following these interviews, a number of clarifications/modifications were made to some tasks, including the development of student worksheets for each task.

- Testing some of the tasks with a colleague's primary school age children, which elicited detailed feedback and comments and examples of the students' work.
- The refinement of task structures to reflect these comments and further judgments from the research team reflecting their expertise and experience.
- A field trial of all tasks across two schools, with multiple classes (3-5) at each level. The research design of field trials is shown in Table 2.1.
- Following phase 1, the tasks were modified before phase 2, based on observations of the task in action, and teacher and student interview comments.

Table 2.1: Research design in field trials (NB: Student age range is from 5 years for Foundation (F) grade, to 11 years for Grade 6. For each grade level there were two tasks; A and B)

School	Trial phase 1		Trial phase 2	
School 1 (4-5 classes at all (F-6) grades)	SIA tasks A	Task refinement	SIA tasks B	Task refinement
School 2 (3 classes at grades 3-5, 1 at grade 6)	SIA tasks B	Task refinement	SIA tasks A	Task refinement

For each of the trial phases the following data were generated:

- Video of a case study teacher's class at each year level;
- Observation/field notes of all classes, including informal conversations with teachers and students;
- Interviews with principal and teachers in focus groups or individually, recorded and transcribed;
- Interviews with focus groups of students, recorded and transcribed;
- Recording and transcription of the research team analysis/review discussions; and
- Analysis of student work samples.

Key features of the updated SIA tasks

As a result of the trialling and refinement of the tasks undertaken by the Deakin research team, the following features of the tasks and their layout were successively developed.

- The inquiry skills targeted by each task were modified as appropriate.
- The original feature of having three skills focused for each task was retained.
- Both tasks in each year level were designed to assess the same three skills.
- Skills were selected to be level appropriate and coordinated across all levels.

Findings Informing the SIA task design/redesign cycle

The trialling process supported our belief in the tasks as being engaging and effective tools for assessing science inquiry skills. Teachers and students were generally positive about the tasks, even while offering constructive feedback that prompted further task refinements. Much of this feedback was regarding the practicalities of the tasks, however, a significant proportion of the refinements made were in response to challenges that redirected our methodological focus. These challenges have been grouped into three categories: 1) practicalities, 2) teaching vs. assessing, and 3) familiarity with inquiry. This section of the chapter will describe some of these challenges and the subsequent task refinements that were made.

The practicalities of implementing the tasks

One of the intentions behind developing the SIAs was to minimise the load placed on teachers while still enabling them to effectively assess science inquiry skills in their classrooms. To us, this looked like: limiting the tasks' duration to 1 hour; using resources/equipment that are readily available, affordable, and easy to use; clear teacher instructions, student worksheets, and assessment rubrics for ease in preparing, implementing, and reporting. The feedback we obtained relating to the practicalities of the SIAs prompted us to simplify the language used and reconsider the number of questions in the student worksheets. We replaced supplementary materials, such as images required by students to complete the task, with clearer, less ambiguous, versions, as requested by teachers.

We also found that teachers were struggling to assess three inquiry skills for all of the students in their class during the set time. As a consequence, they tended to rely on their preconceived notions of some of the students and assess only those they felt they needed to:

> Having said that, sometimes it's, how you said, you got your high flyers. So, I might not spend too long looking at the high flyers. I'd probably be observing those kids that I'm wondering where they're at. *Grade 4 teacher on assessing students during the task.*

Nonetheless, teachers were sometimes surprised by the skills demonstrated by some students during the SIA task:

> I think actually that there are some students that I would not necessarily have pegged as the science thinkers who really surprised me, and I think that that's maybe because they're not given as much opportunity to show their thinking. So I think it kind of showed me that there are some students in here that maybe fly under the radar that we're not as aware of who are really science or mathematically minded, or have a lot of these skills but maybe we just don't notice them as much because as they are in classrooms there are other students who kind of take over or who are so much more willing to put their hands up and really want, like being really active in those conversations so we hear from them all the time. *Grade 1 Teacher*

These findings highlight the insights that such inquiry tasks were offering teachers regarding the capacities of some of their students, that they hadn't been able to observe otherwise. They also led us to adjust the rubric templates to clarify that teachers could choose the number of skills they wanted to assess during the lesson based on their assessment needs.

After observing the tasks in action and reflecting on them with teachers, it also became clear to us that the complexity of the questions in the student worksheets needed to be reduced to limit the tasks to 1-hour, and also to match to the level of the students. Despite the SIAs being designed in accordance with the science inquiry skills outlined in the Australian Curriculum, some aspects of the tasks were too challenging for students, particularly in the younger years, due to disrupted learning during the pandemic. However, it was not practical to update the tasks to the observed level as future cohorts may realign with the skills expected in the curriculum. Instead, the language and skills in the rubrics were refined to clarify the skills demonstrated at levels below the standard.

Navigating between teaching and assessing

Although the SIAs function as assessment tasks for science inquiry skills, we found that it was challenging for teachers to navigate the territory between 'teaching' the skills and 'assessing' the skills. This made it difficult for them to control the extent to which they were scaffolding the experiences and learning of their students, rather than stepping back and prompting students to engage their knowledge and skills independently of the teacher. This challenge was representative of the style of teaching the teachers were accustomed to, where their focus was traditionally on explicitly teaching knowledge and skills 'before' assessing. The quotes below from two different grade 4 teachers demonstrate this challenge:

> So, in terms of what we're assessing, that part would probably have been a little bit off to the side of anything we've actually covered with the kids. So, in my head I was broaching this as more of a teaching thing, I think…And I think that probably was one of my big hurdles…If it was an assessment, I wouldn't have talked as much. *Grade 4 teacher*

> I knew that the students didn't know a lot of the processes and the terminologies, so I had to break down the questions and use explicit teaching, and it was harder to conduct an assessment when we knew that they didn't know how to approach the assessment. So we had to guide them a lot, and I had to really write down ideas and really prompt them and sometimes have to give them the answers as well so then they can. *Grade 4 teacher*

In both quotes above, the teachers justify their need to teach the content and skills required to complete the task as it was new to their students, thus they felt they were unable to treat the task as an assessment. Such examples highlight how the teachers are unused to operating assessments independently of content. An extension to this challenge is the argument presented by teachers that the SIAs needed to align with their unit of work:

> …we originally had chosen this because we were going to be teaching measurement and maths at the same time, so it was going to link in really nicely with our mathematics unit so that was another great plus, and then we had to shift things around and do statistics and probability instead. But that would have been another plus if this science task was done at the same time as our maths unit which was about length and mass and capacity, it would have been a really nice link. *Grade 1 teacher referencing a task that required students to order/rank measurements.*

The quote above from the grade 1 teacher exemplifies this point, emphasising that there is not yet a teaching culture that focuses on assessing inquiry skills independent of content knowledge. Consequently, it became clear to us that there was a need to further emphasise this focus on assessment within the SIA teacher documents. This prompted the development of the SIA Introduction document and rationale, which outlined how to effectively use the SIA tasks and emphasise that they were designed to be used independently of the curriculum units. We also refined the language and structure in the teacher instructions to encourage the prompting of ideas over explicit instruction. Additionally, the language in the rubrics was adjusted to prioritise the assessment of level of student performance independent of teacher support (for instance, with levels distinguished by whether a student could demonstrate an inquiry practice with or without teacher prompting).

Some of the challenges of navigating between teaching and assessing intersect with the difficulties associated with navigating unfamiliar territories of inquiry-based learning. As an example, we found that there were conflicting perceptions of 'inquiry', where teachers' interpretations were not consistent with the investigative practices that formed the SIAs. Some potential consequences of this and other difficulties that arose relating to a lack of familiarity with recognising or assessing inquiry are addressed in the coming section.

(Un)Familiarity with inquiry

The final major set of task modifications were made after finding that teachers were needing greater support in the recognition and scaffolding of higher-level performance. This was because often teachers were unfamiliar with the nature of the inquiry skill/s being assessed. Some teachers also found this difficult as they were inexperienced with teaching and assessing inquiry skills, particularly in the investigative style presented in these SIAs:

> I feel like we as a level jumped at this opportunity because we don't have enough assessment in this inquiry and science stage. *Grade 5 teacher*

> There's not a whole lot of times when the students can work in groups to investigate something. It's very teacher led when we do science work in Year 1 or what we've been doing so far. *Grade 1 Teacher*

> Apart from the difficulty of having a plant growth in one session, this actually fit into the hour–we've been struggling with a lot of the hands-on because it just takes time set-up. *Grade 2 Teacher*

The three quotes above exemplify how teachers were unfamiliar with several different aspects of science inquiry. This came down to the schools not including much science inquiry in their curriculum, or adopting heavily scaffolded styles of teaching science that were teacher-led, or feeling that they were unable to embed inquiry learning into their lessons due to time constraints. This unfamiliarity with inquiry assessment made it difficult for teachers to interpret the rubric:

> I think that the rubric was difficult to understand at the very start. When I read through, I wasn't really sure what I was looking for. So, I think if I am to like have that rubric and I think for everyone to get on the same page, a worked example or something for us to understand that that's a one, that's a two. Even in terms of like when we are listening to the conversation, like, what am I listening for, I wasn't even sure. So, I think, clarity in regards to what they have to do to achieve each number is important. *Grade 3 teacher*

The quote above is just one excerpt from many of the interviews where teachers requested student work samples. The teacher packages were subsequently refined to include work samples of different levels of performance for all tasks. Further interviewing revealed that teachers were looking for explicit guidance:

> We're just like the learners–we're going to produce what you want us to produce, but we need to know what we're going for. *Grade 3 teacher*

Teachers wanted clarity on what to do and what to look for. Ergo, we added annotations to each of the student work samples included in the teacher instruction packs to explain why each work sample was evaluated to represent a particular level of performance. We also further refined the teacher prompts in the teacher instructions so that they almost served as a script; providing specific prompting questions and hints for each stage of the investigation and for each of the questions in the student worksheets.

By undergoing this trialling and interviewing process, the final SIA tasks were considerably refined. By the second round of trials, based on the ease with which they implemented the tasks, teachers seemed more familiar with the inquiry tasks, ("*The tasks were a lot smoother this time round–the kids knew what to expect*" (Grade 3 teacher)), validating the potential of the SIAs to support teachers in shifting to a culture of assessment. The overall reviews of the SIA tasks were positive:

> It was really good. The kids were very engaged because…none of them asked me, 'Can we have recess?' None of them asked me, so they worked

all the way 'til the bell rang which meant that it was really good. *Grade 5 teacher*

We had to do it by ourselves. And after that, we knew what we could do. *Grade 4 student*

Discussion

This research, for us, exposed a fundamental tension between the principles applying to naturalistic assessment settings that attempt to measure competencies in-situ, compared to tightly restricted assessment settings in which the task and assessment protocols can be tightly constrained to maximise reliability of measure. In this chapter we have outlined our argument for an inquiry assessment approach that is situated in naturalistic tasks and scored by the teacher based on field observations as well as student productions. To recapitulate, our justifications for this approach concern both students and teachers.

Concerning students

From the perspective of conducting a valid assessment of student inquiry skills, we believe it important to have students engage productively with the task in order to show their full potential in conducting inquiry. For this, the structured nature of the task and the strategic involvement of the teacher is important, even if this introduces constraints on reliability. In justifying this approach, we draw on the construct of the Zone of Proximal Development (ZPD: Vygotsky, 2012) that distinguishes between performance without, and with the support of an expert other. We are interested in what students are able to do if they understand, are interested in, and are minimally supported to engage with the inquiry task. This implies also that the task is meaningful for the student, and worthy of investment of time and thought. The practical, collaborative setting of the tasks and their everyday contexts were therefore important.

The second point to make from the student perspective is that the 'inquiry skills' as characterised in curricula, will differ in expression depending on the inquiry setting, and are inherently intertwined with some content knowledge. Student questioning, for instance, may differ depending on a student's knowledge of the phenomenon, and will also be influenced by their capacity to observe and notice particular features. Control of variables also applies to ways of achieving reliable measurements and not simply the overall design. Analysing can take many forms from visual representations of science ideas to the construction and interpretation of tables and graphs. These complexities place inevitable constraints on both reliability and validity, which must acknowledge the complexity of science inquiry practices.

Concerning teachers

Part of the context for these tasks concerns the lack of experience of primary school teachers in carrying out such structured science investigations, and in making judgements about the individual inquiry components or skills. Even with teachers who privilege science topics in their practice, we found that focusing on specific inquiry practices was a new experience for them. An important feature of the assessment tasks, therefore, was their educative role in providing models of structured inquiry that was do-able in primary school classrooms, and models of teacher questioning and support that challenged students to develop the inquiry practices. Further to this, given the abstracted nature of the characterisation of inquiry skills, there was a need to interpret these for the specific contexts, and an important part of the research was to develop rubrics that were able to be meaningfully applied in the particular context. In this practical context it was inappropriate for teachers to make judgments simply on students' reports–sometimes reports from a group were very similar yet it was clear the individuals had different levels of involvement. The rubrics included observational judgments of student behaviours and contributions to group processes. Again, this was educative for teachers in clarifying the in-situ interpretation of the particular skill.

The methodological tension

The development of system assessment approaches places great emphasis on the clarity of measurement constructs and the reliability of their measure. Research into producing and validating such instruments customarily proceeds with the design of written items and statistical approaches to validate constructs and the reliability of their measure. In the SIA research, because of the wider framing of the assessment purposes described above, we therefore had to think through the tensions between the validity of the measure and its reliability. For us, given the need to involve teachers in assessment decisions through field observation together with judgments of student productions, and the principle that the assessment tasks model inquiry tasks and the teacher's conduct of these to support learning, the construct of validity was inevitably privileged relative to reliability. It is often claimed, in assessment circles, that you cannot have validity without reliability. For us, a single-minded pursuit of reliability in assessment would have compromised the validity of the approach, for the reasons described above. A valid assessment required active engagement of both students and teachers in generating a common and meaningful purpose for the inquiry practices.

Pursuing this principle engaged us with two methodological challenges:

1. Expanding our notions of validity to recognise its complexity, and adjusting our methods to accommodate this; and
2. Dealing with the complex interconnections between teacher practice and learning, and assessment.

In Table 2.2 we outline the insights generated by the project concerning the conditions for validity of the SIA assessments, and the implication for the research into constructing and validating these.

Table 2.2: Methodological implications relating to validating the SIA tasks

Validity requirement	**Implications for research method**
Students need to be sufficiently engaged with the logic of the task to perform at level, including appropriate supports	Field observations of students responding and being supported, interviews with students
The task needs to be manageable in a practical sense of equipment, time, and strategies for applying the rubric	Field observations, interviews with teachers, and video transcripts for analysis and representation of teacher support instances
Teachers need to understand the nature of the inquiry skills being assessed, including how they play out for the particular task	Field observations, teacher interviews, team review discussions

Teachers need to be committed to assessing inquiry–this involves providing appropriate support that allows students to demonstrate their competencies	Field observations, teacher interviews, team review discussions
The rubrics need to be meaningful to teachers in terms of the nature of the particular skill expressed in the particular task	Professional review, trialling of the wording and its use for assessing, teacher interviews
The rubrics need to allow discrimination between students	Collection and analysis of student work samples and performance descriptions to illustrate levels

As with much research, the methodological emphases in this research changed over the course of the project, in this case because of the complexity of what we were attempting, with the needs for a reliable and valid assessment approach rubbing up against school cultures. Our major aim of refining and validating the SIAs became less about establishing the reliability of the assessments and more about refining the tasks and surrounding resources to support teacher learning that would maximise the validity of the assessments.

We argue that one cannot have a useful and reliable assessment resource for schools unless there is a culture of recognition by teachers of the constructs being assessed. The requirement of 'validity' that the assessment instruments represent professional judgments of the particular constructs being measured, must extend to the judgments of the teachers using the instruments. Thus, the twin goals of this project, to construct valid measures of inquiry practices, and to educate teachers to recognise and support the development of these constructs, are inextricably linked. Validation of instruments like the SIAs must include their refinement to support teachers' growing recognition of the nature and possibilities of articulating and assessing inquiry practices.

Conclusion

The tension between reliability and validity in this research arose from the nature of school cultures around primary science teaching and learning that frame teacher knowledge and practice. Developing in-situ science inquiry assessments inevitably involved supporting teachers to recognise and manage students' inquiry practices as well as learning to make judgments about them. For the research team, the question became: Do we focus on students, or teachers, or both?

We shifted our emphases as we realised the complexities of what we were dealing with; our focus shifted to include an emphasis on clarifying the constructs and establishing supports for teachers to organise and value the assessment of particular skills. This was more important than simply focusing on the refinement of rubrics for reliable assessment that would be of limited validity and influence without the preparation of teacher supports we ended up focusing on.

Overall, the question we asked ourselves is: What approach to inquiry assessment will best support system level improvement in student learning. We believe these inquiry assessments are a step in that direction.

Acknowledgements

The Deakin research team would like to acknowledge the schools–principals, teachers, students, and parents for enabling this research at such a challenging time (Term 4, 2022). The team would also like to acknowledge the PMSS Cohort 6 teachers for their preparedness to support the research around the SIA task research. Finally, thanks to our colleague and her children for trialling many tasks.

References

ACARA: Australian Assessment, Curriculum and Reporting Authority (2022). Australian curriculum: Science. https://v9.australiancurriculum.edu.au/

Anderson, R. (2002). Reforming science teaching: What research says about inquiry. *Journal of Science Teacher Education, 13*(1), 1-12.

Brown, A. L. (1992). Design experiments: Theoretical and methodological challenges in creating complex interventions in classroom settings. *The Journal of the Learning Sciences*, *2*(2), 141-178.

Capps, D. K., & Crawford, B. A. (2013). Inquiry-based instruction and teaching about nature of science: Are they happening? *Journal of Science Teacher Education*, *24*, 497-526.

Carey, S., Evans, R., Honda, M., Jay, E., & Unger, C. (1989). 'An experiment is when you try it and see if it works': A study of grade 7 students' understanding of the construction of scientific knowledge. *International Journal of Science Education, 11*(5), 514-529.

Chen, H-L. S., & Tytler, R. (2017). Inquiry teaching and learning: Forms, approaches, and embedded views within and across cultures. In M. Hackling, J. Ramseger, & H-L. S. Chen (Eds.), *Quality teaching in primary science education: Cross-cultural perspectives* (pp. 93-122). Springer. DOI: 10.1007/978-3-319-44383-6_5

Christensen, K., & West, R. E. (2018). The development of design-based research. In R. E. West (Ed.), *Foundations of learning and instructional design technology* (pp. 483-509). https://edtechbooks.org/pdfs/mobile/lidtfoundations/_lidtfoundations.pdf

Cobb, P., Confrey, J., Di Sessa, A., Lehrer, R., & Schauble, L. (2003). Design experiments in educational research. *Educational Researcher*, *32*(1), 9-13.

Dewey, J. (1933). How we think: *A restatement of the relation of reflective thinking to the educative process*. D.C. Heath.

Dewey, J. (1996). Essays. In L. Hickman (Ed.), *Collected work of John Dewey, 1882-1953: The electronic edition*. InteLex Corporation.

Duggan, S., & Gott, R. (2002). What sort of science education do we really need? *International Journal of Science Education*, *24*(7), 661-679.

Duschl, R. (2008). Science education in three-part harmony: Balancing conceptual, epistemic, and social learning goals. *Review of Research in Education, 32*(1), 268-291.

Edwards, D., & Mercer, N. (1987). *Common knowledge: The development of understanding in the classroom*. Routledge.

Goodrum, D., Druhan, A., & Abbs, J. (2011). *The status and quality of Year 11 and 12 Science in Australian schools.* Australian Academy of Science.

Goodrum, D., Hackling, M., & Rennie, L. (2001). Research report: The status and quality of teaching and learning of science in Australian schools. Canberra; Department of Education, Training and Youth Affairs. http://www.detya.gov.au/schools/publications/index.htm

Holmes, N. G., & Wieman, C. E. (2016). Examining and contrasting the cognitive activities engaged in undergraduate research experiences and lab courses. *Physical Review Physics Education Research*, *12*(2), 020103

Masters, G. N. (2014). Assessment: Getting to the essence. *Designing the Future, 1*, 1-6. https://www.acer.org/files/uploads/Assessment_Getting_to_the_essence.pdf

National Research Council. (2013). Next generation science standards. Retrieved from http://www.nextgenscience.org/three-dimensions

New Zealand Ministry of Education. (n.d.). *Assessment online*. https://assessment.tki.org.nz/

Osborne, J. (2006). Towards a science education for all: The role of ideas, evidence and argument. Proceedings of the ACER conference: Boosting Science Learning - What will it take? ACER.

Palm, T. (2008). Performance assessment and authentic assessment: A conceptual analysis of the literature. *Practical Assessment, Research & Evaluation*, *13*(4), 1-11.

Plomp, T. (2013). Educational design research: An introduction. In T. Plomp & N. Nieveen (Eds). *An introduction to educational design research* (pp. 9-36). Netherlands Institute for Curriculum Development (SLO).

Reeves, T. C. (2006). Design research from a technology perspective. In J. Van den Akker, K. Gravemeijer, S., McKenney, & N. Nieveen (Eds.), *Educational design research* (pp. 52-66). Routledge.

Schwab, J. J. (1962). The teaching of science as enquiry. Harvard University Press.

Schwab, J. J. (1965). Structure of the disciplines: Meanings and significances. In G. W. Ford & L. Pugno (Eds.), *The structure of knowledge and the curriculum*. Rand McNally.

Stillman G., Cheung K., Mason R., Sheffield L., Sriraman B., & Ueno K. (2009) Challenging mathematics: Classroom practices. In P. Taylor & E. Barbeau (Eds.), *Challenging mathematics in and beyond the classroom* (pp. 243-283). Springer. DOI:10.1007/978-0-387-09603-2_8

Tytler, R., Hobbs, L., Brown, J., White, P., Campbell, C., Cripps Clark, J., Delaney, S., Herbert, S., Peters, A., Sawatzki, C., & Widjaja W. (2021). Theory and practice relations in design-based research: Designing professional learning with teachers teaching mathematics and science out-of-field. In P. White, R. Tytler, J. Ferguson, & J. Cripps Clark (Eds.), *Methodological approaches to STEM education research volume 2* (pp. 40-65). Cambridge Scholars Publishing.

Vygotsky, L. S. (2012). *Thought and language*. MIT press.

Russell Tytler (BSc (Hons), BEd, MSc, MEd, PhD, FASSA) - Deakin University, School of Education, 221 Burwood Highway, Burwood, 3125, Victoria, Australia.

Russell Tytler is Alfred Deakin Professor and Chair in Science Education at Deakin University, Melbourne. He has researched and written extensively on student learning and reasoning in science. His interest in the role of representation as a multimodal language for reasoning in science extends to pedagogy and teacher learning. He researches and writes on student engagement with science and mathematics, aesthetics and identity in learning, school-community partnerships, and STEM curriculum policy and practice. He is widely published and has been chief investigator on a range of Australian Research Council and other research projects.
https://orcid.org/0000-0003-0161-7240

Peta J. White (BSc (Hons), MRSysMan, PhD, SFHEA) - Deakin University, School of Education, 221 Burwood Highway, Burwood, 3125, Victoria, Australia.

Peta J. White is an Associate Professor in science and environmental education at Deakin University. She educated in classrooms, coordinated programs, supported curriculum reform, and prepared teachers in several jurisdictions across Canada and Australia. Peta continues her commitment to initial teacher education leading courses, units, and programs and in-service teacher education through research-informed professional learning programs. Peta's current research follows three narratives: science and biology education; sustainability, environmental, and climate change education; and collaborative/activist methodology and research.
https://orcid.org/0000-0002-0225-5934

Shefali Sharma-Wallis (BSc (Hons), MTeach) - Deakin University, School of Education, 221 Burwood Highway, Burwood, 3125, Victoria, Australia.

Shefali Sharma-Wallis is a PhD candidate at Deakin University. Her experiences as a teacher of maths, general science, biology, and STEM have prompted her research interests in science and STEM education, and teacher education. Shefali's PhD project explores the influence of disciplinary teacher identity on the practices and perceptions of integrated STEM teachers.
https://orcid.org/0009-0004-2749-2153

CHAPTER 3

INTEGRATING GUIDED INQUIRY IN SENIOR BIOLOGY SETTINGS: METHODOLOGICAL AFFORDANCES OF TEACHERS AS CORESEARCHERS

AMRITA KAMATH & GARY SIMPSON

Abstract

Design-based research (DBR) is undergirded by teacher-researcher partnership, wherein educational interventions are collaboratively planned and contextually enacted with a common goal to address an issue of significance. In this chapter, a researcher (Amrita Kamath) and a participating teacher (Gary Simpson) describe the affordances of co-research experienced when undertaking a study to investigate the effectiveness of guided inquiry in senior biology settings to enhance student learning and engagement. Collaboration of this nature comes with a set of methodological considerations in terms of theory and practice, and three main challenges encountered in the process are discussed in this chapter: 1) Finding participants/acquiring consent; 2) Navigating pressures of senior secondary; 3) Addressing contextual influences, including school culture and student motivation. Discussing the issues from the perspective of teacher and researcher, the authors clarify their distinct roles, expectations, reflections, and key findings.

Key words: Design based research, Co-research, Guided inquiry, Senior biology

Introduction

Design Based Research (DBR) is a methodological approach that aims to produce new theories and practices by exploring learning processes in naturalistic educational settings (Aksela, 2019; Plomp, 2013). Heralded as a practical research approach that can "bridge the chasm" (Anderson & Shattuck, 2012, p. 16), between educational research and inherent complexities of classroom practice, DBR has potential for rapid pay-off as it is built around issues faced by teachers in the course of their work (Amiel & Reeves, 2008; Unsworth et al., 2021). In contrast to classical experimental methodologies where comparing experimental conditions with a control may lead to over-simplification of classroom complexity, DBR offers opportunity to develop theory focusing on ongoing complex educational practices (Tytler & White, 2022). DBR also has the potential to maximise improvement for learners, as it aims to generate optimal solutions to contextually specific problems (Reeves, 2006).

The purpose of DBR is twofold, it can be adopted to design interventions that validate/develop theories, or provide a solution to a pertinent educational issue (Plomp, 2013). Both aims amalgamated through this research, which sought to address the issue of student disengagement/lack of understanding in senior biology, using guided inquiry, a pedagogical approach with deep theoretical underpinnings. The study, titled "*Effective enactment of guided inquiry in senior biology settings–A critical examination through design-based research*" had two research questions:

- How can guided inquiry be effectively enacted to enhance student learning and engagement in VCE biology?
- What cultural, technical, and political influences shape effective enactment of guided inquiry in VCE settings?

Five teachers and their respective cohorts from four schools (two state and two independent) participated in this study. This chapter focuses on a reflective discussion of the collaboration between one teacher, Dr. Gary Simpson and the researcher, Amrita Kamath, to deeply explore nuances of the methodology. Epistemologically, the perspectives presented in this chapter stem from unique lived experiences, with quality professional conversions enhancing the meaning-making process of the collaborative experience (Conway & Abawi, 2011).

Research design

The systematic, iterative steps adopted for this study are illustrated in Figure 3.1. After the establishment of the research questions, a prototype guided inquiry sequence was designed by the researcher, informed by theory and existing design principles, and drawing on inquiry-oriented contemporary resources that have been created through extensive research (Plomp, 2013; Reeves, 2006; Contemporary Science Practice in Schools, 2019). Data collection was undertaken via semi-structured interviews of teachers, classroom observation and recording, student focus group discussions, student/teacher artefacts and field notes. Braun and Clarke's (2006) steps of thematic analysis were consistently applied to the data; these steps included getting familiar with data, generating initial codes, searching for themes, reviewing themes, defining and naming themes, and using this information as a foundation for discussion and planning of subsequent interventions, and to elicit overall findings. Each step of the process required careful consideration, negotiation, and navigation by the teachers and researcher, as described in the subsequent paragraphs.

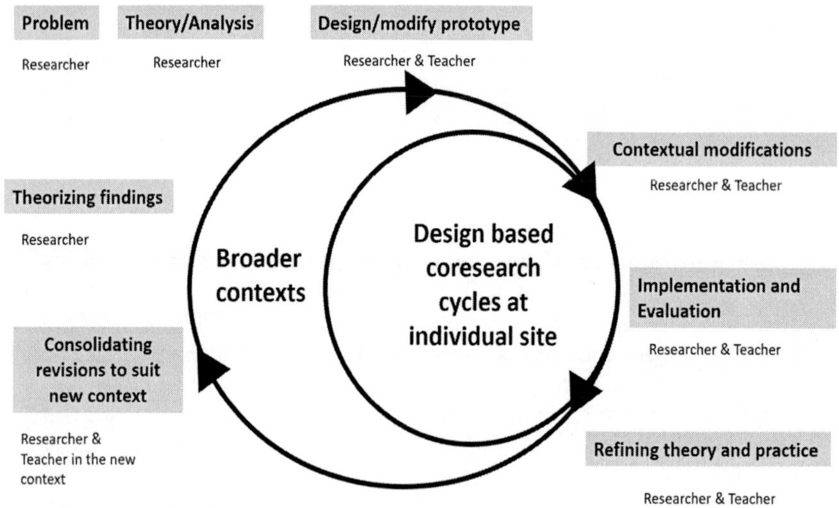

Figure 3.1: Steps of DBR indicating distinct and collaborative roles of the teacher and researcher (Adapted from Tytler & White, 2022, p. 145)

Methods of data collection

Semi structured interviews with the teacher

Semi-structured interviews with the teacher were conducted and recorded at two stages–one before and one after implementation of interventions. The first interview afforded opportunity to collect data on the teacher's identity, experiences, belief systems, context encompassing school culture and climate, and the influence of both these elements on the teacher's perceptions and practices of guided inquiry. Insights gathered via post-intervention interview focused on evaluating and analysing the teaching and learning processes, and developing suitable adaptations to the lesson sequence through reflections on the experience. Questions designed for the semi-structured interview were open ended and focused, not leading in any way, with probes provided to purposefully guide the conversation (Creswell & Creswell, 2018).

Teacher-researcher collaborative sessions

DBR is deeply rooted in collaboration between practitioners and researchers (Conway & Abawi, 2011). Two formal sessions were conducted, one before the interventions to collaboratively infuse contextual modifications into the sequence and one midway, after four lessons of the eight lesson sequence had been implemented. The sessions were video recorded to capture teacher insights, reflections of dialogic teaching, and possible ways of further refining the pedagogical design of the teaching learning sequence to better fit the technical, cultural, and political dimensions of their context (Fang, 2021).

Artefacts

Students produced artefacts and a record of their adjustments and changes throughout lessons was documented to give greater insight into the productivity of interventions. Student work was collected at the conclusion of each lesson and photographed, with original artefacts returned to students as soon as possible. Classroom board work was photographed during and at the end of each lesson to capture and record co-construction and sharing of ideas, thinking, and understanding. Students and teacher were encouraged to share teaching/learning experiences and thought processes based on the artefacts.

Classroom recording and observations

Lessons were videoed by the researcher from within the classroom, ensuring steps to minimise disruption like being at an unobtrusive spot and placing the cameras in positions/angles that were least distracting. Key classroom proceedings were also documented as field notes by the researcher. These observations were used as probes for reflection during collaborative discussions, and informed adaptations to the sequence before subsequent iterations.

Student focus groups

Student voice was respected, enabled, and supported throughout this study. Two formal 30-minute focus group discussions were conducted with consenting students (process of procuring consent described on page 54), one at commencement and one after the interventions. Students also shared their thoughts and feedback through informal discussion with the teacher and researcher, and this anecdotal evidence proved a rich source of student insight.

Methodological elements: Experience and reflections

Coresearching: Clarifying roles of the researcher & teacher

Coresearch, as defined by Barton (1992, p. 114) “is a process which revolves around the interactions between people operating in a particular social and cultural context, where the researcher is simultaneously a participant and an observer”. Establishing clarity of insider-outsider positioning of the researcher and teacher is critical in the process of coresearch (Mariguddi, 2022). An insider shares the characteristic, role, or experience with the participants whereas outsiders lie beyond the commonality shared by participants (Dwyer & Buckle, 2009). Considering this definition, the researcher fits as an insider based on their qualification and experiences as a secondary school teacher, and an outsider to the setup and its contextual influences. The participating teachers could be termed as complete insiders in their respective schools. This positioning and reflexivity has offered immense opportunity to draw nuanced connections into classroom discourse and practices, and maintain a critical awareness of its possible influences on the process (Unluer, 2012).

To ensure procedural clarity, it is vital to identify the distinct role and responsibility of the key operatives of the process, in this case the researcher and the teacher. Table 3.1 lists the roles linked to each step of the process.

Table 3.1: Articulation of distinct roles of researcher and teacher in the process of coresearch

Step	Role	Description
Identification of the Problem	Researcher	Problem was established through literature and research.
Theory/Analysis	Researcher	Problem was analysed through literature to formulate appropriate research questions.
Design/modify prototype	Researcher/Teacher	Prototype guided inquiry lesson sequence was developed by the researcher, which was presented to interested teachers.
Contextual modifications	Researcher/Teacher	The sequence was contextually modified based on discussion with teacher to suit their respective cohorts.
Implementation and Evaluation	Researcher/Teacher	Teacher implemented the sequence, with researcher observing and recording classroom practice.
Refining theory and practice	Researcher/Teacher	Collaborative discussions throughout implementation (after lessons) to reflect on interactions/effectiveness and make consistent updates to sequence. Student voice also considered.

Consolidating revisions to suit new context	Researcher/Teacher	The refined sequence and findings taken to three new contexts, with core steps repeated with new teachers.
Theorising findings	Researcher	Findings from the four sites consolidated for theory-practice based solutions to the problem.

Amrita's reflections

As a researcher with prior experience as a secondary science teacher, it is vital that I recognise my role in the study. This reflexivity enables me to appreciate the influence of my experience-based belief system on the process, while also acknowledging personal motivation for this research (Terry et al., 2017). Born and brought up in Mumbai, India, I studied in a highly competitive educational environment. The number of applicants for seats in coveted higher education institutions in India far outweighs availability, and results of senior year examinations and standardised tests serve as determinants to future opportunities. This rigid selection process has led to a singular focus on academics, with the high stakes and technical complexity creating pressures that leave limited scope to move beyond transmissive teaching (Cheney et al., 2005; Kumari, 2016).

However, there are always exceptional educators who, despite the challenges, strive to motivate their students. This narrative would be incomplete without expressing my gratitude to my High School Biology teacher, who adopted pedagogical innovations, investing additional time and effort to kindle a curiosity for learning science in my young, impressionable mind. After high school, I pursued a Bachelor of Science (BSc–Life Sciences) then a Master of Science (MSc-Applied Medical Sciences). Experience as a researcher in a genetic laboratory enhanced my skills in scientific research, but I was keen on a career that amalgamated two fields I am passionate about biology and education. So, I decided to study further, completing a Bachelor of Education (BEd).

My teaching journey began in India, after which I relocated to Bangkok, Thailand, where I continued to teach secondary science at international schools for eight years. While in Bangkok, alongside teaching, I earned a Master of Education (MEd) designed for in-service teachers. I also participated in educational research as a practitioner in my middle school principal's doctoral project titled *Teacher evaluation that matters: A participatory process for growth and development* (Paryani, 2019). This study was undertaken as three action research cycles over a period of 1.5 years, where a team of six teachers (including myself) worked closely with the principal to develop a contextually relevant teacher evaluation model. Through this experience, I developed a keen interest in educational research, and recognised the value of the voice, cultural disposition, and practices that engender resilience for teachers.

I then relocated to Australia, where, in order to integrate into the Australian education system and qualify for teacher' registration, I completed a further teaching qualification, Master of Teaching (MTeach). As my professional experience and outlook broadened, I became further inspired to enhance my pedagogical repository, and my search took me to the affordances of Inquiry Based Learning (IBL). I conducted a phenomenological study of senior biology teachers' perceptions and classroom practices of IBL (Kamath & White, 2023) findings of which established the theoretical and conceptual foundation of this research. The research revealed that senior biology teachers experienced tensions in teaching through inquiry and felt compelled to adopt transmissive approaches, despite being in favour of IBL.

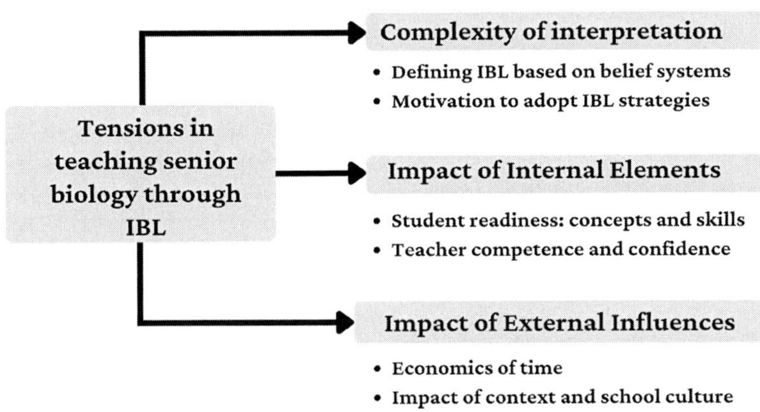

Figure 3.2: Tensions in teaching senior biology through IBL (adapted from Kamath & White, 2023)

Teachers' Role

As described by Unsworth et al. (2021):

> …DBR relies on teachers' preparedness to be active contributors to the cycle of intervention design, implementation, evaluation and revision. This kind of collaboration is distinctive to DBR and very different from the

> participant collaboration that is involved in say, ethnographic, observational or even quasi-experimental research. (p. 67)

With this methodological uniqueness comes a distinct set of requirements in terms of time, and flexibility. Guided inquiry based strategies are learner-centric, wherein students are stimulated to ask authentic questions and think critically, at progressively higher levels of cognition. To facilitate purposeful conversation and learning, the teachers' role broadens to that of a guide, a motivator, and an innovator (Spronken-Smith, 2012). Designing and implementing individualised scaffolds and differentiation techniques is also imperative to consolidate understanding (Donohue et al., 2020). These expectations, combined with the interventionist nature of this research, required a strong commitment from the teachers in terms of time and motivation.

Preparedness to enact innovative teaching and learning strategies is shown to be directly linked to teachers' self efficacy and pedagogical content knowledge (Han-Tosunoglu & Lederman, 2021). Self efficacy is defined as a teacher's personal judgement of capability, and pedagogical content knowledge, describes a teacher's awareness about how particular content can be taught in ways that enhance student understanding (Loughran et al., 2012). Both these terms are synergistic, and determined by multiple factors including identity, belief systems, experience, and external elements like school culture and policies (Fang, 2021). As self efficacy and pedagogical content knowledge is deeply impacted by a teacher's background and context, the following section describes what it meant for the participating teacher.

Gary's reflections

> I have been teaching Science and Biology for 35 years and have completed a number of postgraduate degrees that were highly reflexive and centred on my own classroom practice. These include a Master of Education and a Master of Arts in Science, Technology and Society, both at Deakin University and culminating in a Doctor of Philosophy in Science Education from Curtin University where I worked with the educational transformist, Professor Peter Taylor. I identify as having high pedagogical content knowledge (PCK) with strong content knowledge, strong pedagogical knowledge and, as a critical constructivist, one with a strong desire to develop student agency through sharing control of the teaching and learning process.

I prefer to view education, even at the senior year level, as an opportunity to transform understanding rather than transmit knowledge. I clearly display confidence in my abilities in the classroom and a sense of self-efficacy, whilst still being keen to learn and develop my skills. Hence, my enthusiasm to join the project and enter into a collaborative, co-research relationship with the researcher. I can also identify with the outcomes of Amrita's earlier research, quoted above. As someone keen to use inquiry methods with my senior students, while still needing to prepare them for an external exam, the tension between transmission of knowledge and a desire to create students skilled in inquiry for biological research beyond the end of Year 12 exam has been a constant.

As I'm always keen to learn and develop my knowledge and skills and I had an area of study addressing how humans use science to explore and communicate contemporary bioethical issues to teach for the first time, the opportunity to work with Amrita looked like a win-win for me. This would help me to devise an engaging and informative teaching and learning sequence, offer me an opportunity to learn some new skills and knowledge, and help a colleague on their own professional learning journey. As a teacher who has been working with inquiry learning and problem-based learning approaches in an attempt to engage students in their learning by giving them agency and sharing control of decision-making, I was well prepared for trialing this particular innovation. It also meant that my students were prepared to participate in the process as they had experience with structured, guided, and open inquiry models.

The researcher had developed a detailed and comprehensive eight-lesson sequence that effectively guided the students through the inquiry, introducing the students to ethical principles and guidelines with topics that linked clearly to their studies. We were then able to meet and tweak the materials prior to the sequence commencing and prior to each lesson, reflecting on each lesson and looking to meet the various needs of the students in the two classes. We were also able to involve the students in some of the decision-making around the management of the teaching sequence. This was really important to me, as the scaffold was strong enough to achieve the desired result and flexible enough to deal with the individual needs of each class and the different requirements of each student.

Dimensions of challenge to coresearching

In this study, three main challenge areas were encountered and addressed through collaborative discussions between researcher and teacher. It is crucial to state here that though distinctly stated for analytical clarity, these aspects are interrelated.

- Finding participants and acquiring consent.
- Navigating the pressures of senior secondary settings.
- Addressing contextual influences: School culture and student motivation.

Finding participants/consent for the study

This study sought to recruit up to six teachers currently teaching year 11 biology (along with their respective cohorts) from schools across Victoria. There are two main types of schools in Victoria, government and non-government. The key distinctions are that state schools' administration and funding is controlled by state and territory governments, are zoned, and students attend at no or minimal cost. Non-government schools, which can be independent or faith based schools (including Catholic, Muslim, and Jewish) are not part of the government system and have their own fee structure and enrolment stipulations while also receiving possible public funding through the federal and state governments (State Government of Victoria, 2022). Parents and students select schools based on multiple factors, including ranking, proximity, and financial capacity. Cultural factors such as religious beliefs or socio-political perspectives also determine parents' support of faith based or secular schools. These factors determine school culture (further discussed on page 58) and impact the ability/will for involvement in research initiatives. The following sections explicates the recruitment process from the perspective of the researcher, and the teacher.

Amrita's experience

The first, and probably the most significant, challenge was to find participants for this study. After acquiring ethics approval, an email with all relevant information regarding potential benefits and risks, participants' rights, confidentiality of data and dissemination of results was sent through professional networks, inviting practicing senior biology teachers in Melbourne to participate. A perceptible trend in the expression of interest received from teachers was that the majority of them were from independent schools. Teachers who responded to the invitation email were invited to have zoom/phone conversations to discuss further. At this point, many teachers indicated that they would not be able to sign up, with time constraints being cited as the primary concern.

Here, I must acknowledge that the drop out was not surprising, considering the high pressure environment of senior secondary, the intricate nature of the research design (Figure 3.1) and COVID-19 scenarios. However, this project was part of my doctorate, I had a fixed timeframe, and lack of participants was certainly concerning. Another factor that appeared to have impacted motivation to participate was the timing of this research in terms of the COVID-19 pandemic. In 2022, schools were just returning to face-to-face routines after remote learning due to lockdowns, and strict policies were in place around research in schools (State Government of Victoria, 2022). These restrictions, amalgamating with the already existing pressures of senior secondary (Unsworth et al., 2021) prolonged the process of recruitment. After the zoom meeting, if the teacher indicated interest to participate, an email with details and plain language statement had to be sent to procure administrative/principal consent, and this step took time. In many cases, the teacher had to personally go to the principal and request them to peruse the document and sign off if suitable. Six months of persistent networking and discussions resulted in five teachers from four schools (two state and two independent) signing up.

Students of the participating teachers were given the opportunity to be part of the study (deidentified). Students (with separate consent from parents/guardians) could select from four levels of participation as listed below:

- Highest level: Agree to be in whole class video-recorded lessons, participate in two video recorded focus group of 30 minutes duration, and consent to class work being copied (with name removed).
- Medium Level: Agree to be in whole class video-recorded lessons, and class work being copied (with name removed).
- Low Level: Agree for class work to be copied (with name removed) for research purposes.
- No Participation: Do not consent to participate.

In Gary's specific context, he responded to an invite sent through a professional network, and decided to be involved in the project after a zoom meeting with me to understand the details. He demonstrated deep epistemological and pedagogical alignments to the project, and also undertook the responsibility of getting administrative, parental, and student consent prior to commencement.

Gary's experience

The school currently has a number of collaborations with tertiary institutions (for example, the New Metrics program at Melbourne University and as the Asia-Pacific hub of the Compassionate Systems project at Massachusetts Institute of Technology–both of which I am participating in) and as an institution we are open and ready to participate in projects that can be seen to be of benefit to the professional learning of staff and/or the learning of students. To gain permission to participate from the principal was a straightforward process, who enthusiastically agreed to the project.

Fortunately, the parent and student approval process at my school was also straightforward. I had made the students aware that we would be working with a PhD researcher well in advance and kept talking it up. They were enthusiastic to learn more and signed up eagerly. I took the opportunity to promote this in parent/student/teacher interviews during the previous term and the parents were keen to know more about the research and were highly supportive. The school has an online parent permission process that made it very easy to communicate all the details related to the project and for the parents to sign off on their consent. A number of parents contacted me to check details, but more through interest in the outcomes of the project rather than concerns about their students' privacy or confidentiality.

Navigating the pressures of senior secondary settings

Senior secondary environments are complex and intricate, with pressures of time and technical nature of senior science knowledge restricting scope for pedagogical innovation (Cobanoglu, 2021; Unsworth et al., 2021). Complicated assessment terminology increases semiotic demands, with teachers expected to familiarise students with language that is not common in everyday experiences (Prain & Tytler, 2012). Furthermore, rigid curricular objectives and expectations to perform well against a standardised level of achievement, further limit flexibility in terms of individualised teaching and assessment (Kang & Keinonen, 2018). School context determined the extent and impact of each of these factors, and the subsequent sections discuss the experiential nuances in the context of Gary's school.

Gary's experience

This school is a co-educational Independent K-12 school with three campuses on the Mornington Peninsula. The families are generally of a higher than average socio-economic level with many parents having completed at least one tertiary degree and many with one or more postgraduate degrees. The expectations of the parents and the students is for high levels of engagement in their learning and high levels of academic success. However, the school is also seen as an alternative to the mainstream with a focus on the well-being and pastoral care of all students and a desire for them to develop the 21st century capabilities as well as academic success. While the school is required to teach to the externally mandated curriculum, VCE in the case of this study, programs are individualised or personalised to cater for the widely different needs and abilities of the students.

I do not use text books preferring to develop my own materials in response to student needs and interests and often involving the students in producing those materials. As a critical constructivist I encourage the students to see the Western science tradition as one young, ever changing, and developing tradition of scientific knowledge. In this way, I hope that my students will recognise that science isn't all known and that they can have a meaningful career in science questioning the knowledge that we take as given and developing new understanding of our natural world and lived experience.

The teaching and learning sequence fitted with my epistemological views and pedagogical practices as it was developed as a scaffold to guide the inquiry, while giving agency to the students to modify the lesson content to suit their own interests and viewed the experience of science as a living breathing human activity. The sequence also had a range of different thinking routines and learning activities to engage a range of different learners and caused the students to test what they understood about both the ethical principles and the biological concepts they had been studying.

The flexibility of the scaffold was sorely tested. Having planned for eight 60-minute lessons, both classes were reduced to seven lessons before we had started as I had forgotten about a planned parent/student/teacher interview day that had been scheduled for the end of the first week. The two classes were also very different. One class was quiet, studious, not ready to question the teacher's authority, although this experience had them much more involved in discussing what we were doing and why and taking opportunities to question what may come next. They received the full seven lessons and used the time well to develop richly detailed and at times sophisticated responses to the guiding questions and prompts. While the other class was much more robust, less academic, but much more interested in sharing control of the teaching and learning process and telling us what they were thinking. The second class also had some impacts from illness of significant student leaders during the sequence and we discovered that two students would not be present for the final class (they were going on holidays early) so we had to squeeze the sequence into six lessons. But they were still able to successfully display what they had learnt. Both groups achieved great success as evidenced by their final performance and by the various tasks they submitted during the sequence. They were also highly motivated and engaged with the process.

Amrita's experience

The data collection plan at Gary's school required constant adaptation, and as described above, flexibility was key in effective scheduling. Gary's expertise, combined with his rapport with students facilitated purposeful planning. For instance, when the date of the summative task (a stakeholder role play to be performed in the classroom) was announced, students who planned to go on holiday were forthcoming and shared that they would not be available. This is a classroom reality, and although it may not have been ideal to condense the length of the sequence, we were able to collaboratively modify it to ensure all students could participate, and were assessed fairly. Also, as Gary's was the first site where this sequence was implemented, I got insights on how to refine the lessons for future iterations. In the subsequent sites, students were offered the opportunity to present the role play as a podcast, or a pre-recorded representation, circumventing the issue of absenteeism.

Addressing contextual influences: School culture and student motivation

Teaching and learning is an iterative process that goes beyond classroom discourse, and is impacted by a myriad of social components (Thumlert et al., 2018). The shared values, practices, and beliefs in a system represent its culture, and is determined by all stakeholders (Gilbert, 2012). These stakeholders could be intrinsic to the system, i.e. the teachers, students, and administration or be extrinsically involved in the working and decisions, as parents, guardians and policy/curriculum/government representatives. A school culture that encourages teachers to critically reflect on practice and pedagogically innovate is shown to have a positive impact on student motivation and learning (Dobber et al., 2017).

Gary's experience

In the context of this study, the school culture was part of the reason for the success of recruiting the students and their parents and the positive manner in which they accepted the researcher into their class. Both classes were small (n = 9) with a very strong relationship with me, something I try to build with each of my classes in an attempt to develop a community of learners (Palmer, 1993). This could have translated to a distrust of the outsider, but I had prepared the students for the intervention and the researcher was warm and open from the start. The students are also used to student teachers and visitors from outside the school regularly appearing in their classes. It also meant that they were open to try something different and to speak/perform in front of their peers.

Both classes actively took part in class discussions and thoroughly enjoyed themselves with the role plays. This is because the culture of the school encourages such activity in the curriculum and in the rich co-curricular opportunities of the school from Year 1. The manner in which Amrita warmly reached out to the students, using all her experience gained from the classroom in her earlier career, was also a factor in her quick acceptance into both of my quite different classes.

The school has a focus on centering the educational experience on each and every student. We work hard to differentiate and personalise the classroom experience for each child and have extensive opportunities for that in our diverse co-curricular experiences, extensive camping program, and a highly effective pastoral care program that seeks to build a community of learners within their year levels and between year levels. We deal with adolescents, so our aspirations aren't always achieved. However, this context helps to explain how well the students identify with their teacher and felt comfortable in speaking out and sharing their thoughts. They know that their views, their voice, is honoured and that they can speak and they will be listened to. It also helps to explain their motivation to help trial the teaching and learning sequence and participate in the research of a person they didn't know. In fact, they quickly accepted Amrita as part of their classroom team and would engage her in conversation to get to know her and her work better.

Amrita's narrative

"I think it needs a teacher like doc (Gary), someone who is engaging, who is caring, is genuinely beneficial to everyone in the learning process" – Gary's student.

The above statement encapsulates the essence of Gary's influence on his students, and more broadly, the impact of school culture on students' perception of learning. There was the perfect balance between friendly and fair in all interactions between Gary and his students, evidencing the prioritisation of learners. As a reseacher, I was welcomed not only by Gary's students but also by the other staff members, and my interactions with them provided multiple contextual insights. Overall, productive outcomes of this co-research were deeply connected to the receptivity of all the stakeholders towards this research initiative.

Conclusion

This chapter focused on one collaborative experience between a teacher and researcher that occurred as part of a study to explore effectiveness of guided inquiry in senior secondary biology settings. This DBR was set in situ, with the researcher being in the year 11 classroom with the teacher, observing and recording classroom interactions. These observations were then collaboratively unpacked with the teacher in an iterative design process, to refine the guided inquiry based teaching and learning sequence, by identifying inhibiting and enabling pedagogy, monitoring and assessment practices.

The findings at this site demonstrated rapid pay offs, as the researcher was able to apply them to subsequent sites to enhance effectiveness of the guided inquiry sequence (Amiel & Reeves, 2008). Conclusively, a personalised approach to the teacher-researcher partnership (Unsworth et al., 2021) led to productive outcomes, with insights into syncing theoretical underpinning of guided inquiry with realities of classroom practice.

References

Aksela, M. (2019). Towards student-centred solutions and pedagogical innovations in science education through co-design approach within Design-based research. *LUMAT: International Journal on Math, Science and Technology Education, 7*(3), 113-139.

Amiel, T., & Reeves, T. C. (2008). Design-based research and educational technology: Rethinking technology and the research agenda. *Journal of Educational Technology & Society*, *11*(4), 29-40.

Anderson, T., & Shattuck, J. (2012). Design-based research: A decade of progress in education research? *Educational Researcher*, *41*(1), 16-25.

Barton, B. M. (1992). *An evaluation of 'teacher as co-researcher' as a methodology for staff development* (unpublished Master of Education (Hons.) thesis), Centre for Studies in Literacy, University of Wollongong.

Braun, V., & Clarke, V. (2006). Using thematic analysis in psychology. *Qualitative Research in Psychology, 3*(2), 77-101.

Cheney, G. R., Ruzzi, B. B., & Muralidharan, K. (2005). A profile of the Indian education system. Prepared for the *New Commission on the Skills of the American Workforce,* 228-253.

Cobanoglu, N. (2021). The relationship between shared leadership, employee empowerment and innovativeness in primary schools: A structural equation modeling. *European Journal of Educational Research, 10*(1), 327-339.

Contemporary Science Practice in Schools. (2019). *Senior Secondary Science.*https://blogs.deakin.edu.au/contemporary-science-practice-in-schools/

Conway, J. M., & Abawi, L. (2011). Effective cluster collaborations: Transformation through school and university connections. In L. Abawi, J. Conway, & R. Henderson, R. (Eds.), *Creating connections in teaching and learning* (p. 165). IAP.

Creswell, W., & Creswell, D. (2018). *Research design: Qualitative, quantitative, and mixed methods approaches (5th Ed.).* Los Angeles: SAGE.

Dobber, M., Zwart, R., Tanis, M., & van Oers, B. (2017). Literature review: The role of the teacher in inquiry-based education. *Educational Research Review 22 (Supplement C),* 194-214. doi:10.1016/2017.09.002

Donohue, K., Buck, A., & Akerson, V. (2020). Where's the science? Exploring a new science teacher educator's theoretical and practical understanding of scientific inquiry. *International Journal of Research in Education and Science*, *6*(1), 1-13.

Dwyer, S. C., & Buckle, J. L. (2009). The space between: On being an insider-outsider in qualitative research. *International Journal of Qualitative Methods*, *8*(1), 54-63.

Fang, S. C. (2021). Towards scientific inquiry in secondary earth science classrooms: Opportunities and realities. *International Journal of Science and Mathematics Education*, *19*(4), 771-792.

Gilbert, A. (2012). *Courageous pedagogy: Enacting critical science education*. Charlotte, NC: Information Age Publishing.

Han-Tosunoglu, C., & Lederman, N. G. (2021). Developing an instrument to assess pedagogical content knowledge for biological socioscientific issues. *Teaching and Teacher Education*, *97*, 103217.

Kamath, A., & White, P. J. (2023). Tensions in teaching senior biology using inquiry based learning. *Teaching Science*, *69*(1), 29-40.

Kang, J., & Keinonen, T. (2018). The effect of student-centered approaches on students' interest and achievement in science: Relevant topic-based, open and guided inquiry-based, and discussion-based approaches. *Research in Science Education*, *48*(4), 865-885.

Kumari, J. (2016). Public–private partnerships in education: An analysis with special reference to Indian school education system. *International Journal of Educational Development*, *47*, 47-53.

Loughran, J., Berry, A., & Mulhall, P. (2012). Pedagogical content knowledge. In J. Loughran, A. Berry, & P. Mulhall (Eds.), *Understanding and developing science teachers' pedagogical content knowledge* (pp. 7-14). Rotterdam: SensePublishers.

Mariguddi, A. (2022). Co-researching OR researching with teachers in the space between: Pitfalls and benefits. *International Journal of Research & Method in Education*, *45*(1), 86-98.

Palmer, P.J. (1993) *To Know as we are Known: Education as a Spiritual Journey*, Harper and Row, San Francisco

Paryani, P. (2019). Teacher evaluation that matters: A participatory process for growth and development. *The Scholarship.*

Plomp, T. (2013). Educational design research: An introduction. *Educational Design Research*, 11-50.

Prain, V., & Tytler, R. (2012). Learning through constructing representations in science: A framework of representational construction affordances. *International Journal of Science Education, 34*(17), 2751-2773.

Reeves, T. (2006). Design research from a technology perspective. *Educational Design Research* (pp. 64-78). Routledge.

Spronken-Smith, R. (2012). Experiencing the process of knowledge creation: The nature and use of inquiry-based learning in higher education. *International Colloquium on Practices for Academic Inquiry* (pp. 1-17). University of Otago.

State Government of Victoria. (2022). *Conduct research with the Department of Education.* https://www.vic.gov.au/conduct-research-department-education

State Government of Victoria. (2022). *Types of Schools.* https://www.vic.gov.au/types-schools

Terry, G., Hayfield, N., Clarke, V., & Braun, V. (2017). Thematic analysis. In C. Willig & W. Rogers (Eds.), *The SAGE handbook of qualitative research in psychology* (pp. 17-36). London: SAGE Publications.

Thumlert, K., Owston, R., & Malhotra, T. (2018). Transforming school culture through inquiry-driven learning and iPads. *Journal of Professional Capital and Community, 3*(2), 79-96.

Tytler, R., & White, P. (2022). Working with teachers in reform initiatives. In P. J. White, R. Tytler, J. Ferguson, & J. Cripps Clark (Eds.), *Methodological approaches to STEM education research (Vol. 3).* Cambridge, UK: Cambridge Scholars Publishing.

Unluer, S. (2012) Being an insider researcher when conducting a case study research. *The Qualitative Report, 17*(58), 1-14.

Unsworh, L., Tytler, R., & Fenwick, L. (2021). Methodologcial challenges for collaborative research in senior science. In P. J. White, R. Tytler, J. Ferguson, & J. Cripps Clark (Eds.), *Methodological approaches to STEM education research (Vol. 2).* Cambridge, UK: Cambridge Scholars Publishing.

Amrita Kamath (BSc, BEd, MSc (India), MEd (U.S.A), MTeach (Australia)) - Deakin University, School of Education, 221 Burwood Highway, Burwood, 3125, Victoria, Australia.

Amrita Kamath is a PhD candidate and academic at Deakin University. Amrita's experiences include sessional lecturing in the Faculty of Science Education, secondary science teaching, curriculum development, and research in genetics. In educational research, Amrita's current focus is exploring effective implementation of student centric pedagogical strategies in senior secondary contexts, using guided inquiry.
https://orcid.org/0000-0002-0082-2704

Gary Simpson (BEd (Environmental Science), MEd, MA (Science, Technology and Society), PhD) - Woodleigh School, 485 Golf Links Rd, Langwarrin South, 3911, Victoria, Australia.

Gary Simpson is currently the Director of the Brian Henderson Wildlife Reserve, Woodleigh School and teaches Biology and Science. He received the inaugural Teacher Leader Award of the Australian Council of Educational Leaders in 2007, which resulted in an invitation by the Singaporean Ministry of Education to deliver a keynote address to their annual teacher conference the following year. In 2015 he was inducted into the Zoos Victoria Hall of Fame in recognition of his many years working with Zoos Victoria's Education Service and many other conservation reserves and agencies to engage students in conservation of ecosystems and endangered species. Gary recently encapsulated his epistemological and pedagogical approach to teaching and learning in his book "To Teach as an Act of Love".
https://orcid.org/0009-0001-7576-8458

CHAPTER 4

AFFORDANCES FOR LEARNING TO TEACH MATHEMATICS: THE CASE OF OUT-OF-FIELD TEACHERS

JILL P. BROWN & CARLY SAWATZKI

Abstract

Educational theory and the wisdom of practice informed the design of a bespoke Graduate Certificate of Secondary Mathematics for out-of-field teachers of mathematics. Taking a design-based approach, we describe key elements of our design as we traversed the iterative cycle: design → implement → evaluate → refine. This chapter details our focus on the use of digital technologies for the teaching and learning of secondary mathematics. The design of the course was intended to be face-to-face, but due to the global pandemic it had to be changed to a fully online implementation in the first year. This created additional opportunities to model and experiment with a range of digital tools. A framework of affordances is used to describe the opportunities available to students in the course. Our focus is on *Knowledge for teaching in Years 7-10: Statistics and probability* as this second semester unit was in the second cycle of our design.

Key words: Affordances, Digital technologies, Out-of-field teachers, Teaching mathematics, Teacher education

Introduction

This chapter reports Deakin University research involving out-of-field teachers of Year 7-10 mathematics. The authors were part of a team contracted to design postgraduate courses as part of the Department of Education and Training Victoria's 'Secondary Mathematics and Science

Initiative'. By out-of-field (OOF) we mean teachers currently teaching mathematics in Year 7-10 in Victorian secondary schools, with no formal qualification to do so. This generally refers to secondary teachers who have not studied the discipline at university level, nor studied curriculum and pedagogy units focused on the specific teaching of that discipline (e.g., Hobbs, 2013). The project applied educational design research methodology (see Tytler et al., 2021) to create Graduate Certificates of Secondary Mathematics and Science for out-of-field teachers and enable researching of the teaching and learning experience. This chapter sets out to describe the key design elements of the Graduate Certificate of Secondary Mathematics in particular. The curriculum context is the Victorian Curriculum which includes a direct focus on mathematics but also, across all disciplines, on the use of digital technologies and other capabilities such as numeracy and critical and creative thinking (e.g., thinking processes, learning dispositions, metacognition).

The research problem was the need to design a face-to-face course for out-of-field teachers, clearly drawing on our knowledge, experience, and expertise in initial and in-service mathematics teacher education. The intention was to shift the teachers along the continuum towards being in-field teachers. Note, the classification out-of-field versus in-field is a continuum and not a dichotomy. 'Out-of-fieldness' is a multi-faceted concept. A global pandemic shifted the delivery online. The Gibsonian construct of affordances (Gibson, 1977, 1979) underpins the design-based Graduate Certificate(s) as we seek to change mathematics teaching practices.

Digital technologies

Digital technology use is increasingly common in the teaching and learning of secondary mathematics (e.g., Clark-Wilson et al., 2020). It is well established that there is great potential for "the use of digital technologies [to] improve teaching and learning" (Brown, 2015, p. 112). However, we still need to learn more as to how, and why, they are used in order to increase this use in school classrooms (Roschelle et al., 2008) as the effectiveness and impact of this use on learning varies.

Tytler et al. (2021) provide an overview of the project and describe the two courses (one for teachers of mathematics and one for teachers of science). The intended design of the Graduate Certificate of Secondary Mathematics (GCSM) included five days of face-to-face classes for each of the four units in the course. At Deakin University (and most other universities) all units are supported by online components (e.g., websites including digital content, discussion spaces, email). However, the global pandemic necessitated a rapid pivot to online classes at universities and schools across the state of Victoria (and elsewhere). The design phase of this research project included a significant timeframe during which the designers were unsure if all four units in the course would be face-to-face as intended or online. Critically, the lockdowns increased the need to ensure high quality delivery via online modes. Having out-of-field teachers as our students raised some very specific challenges. Previously, Brown et al. (2022) presented an analysis of aspects of the course design and the tensions between theory and practice involved. In the research reported here, two broad categories of digital technologies are of interest, namely *mathematical digital technologies* and *digital technologies for teaching and learning*.

Mathematical digital technologies

The term mathematical digital technologies, refers to technologies specifically designed for ‘doing mathematics’, or those that can be used to do mathematics. Computer Algebra System (CAS) enabled graphing calculators that have long been used in Victorian secondary schools (Leigh-Lancaster, 2003; Leigh-Lancaster & Stacey, 2022; VCAA, 2005, 2016). Graphing calculators were allowed in high-stakes examinations from 1997 and assumed to be widely in use in mathematics classrooms from 1999. CAS-enabled graphing calculators were piloted in examinations from 2002 and routinely used for almost two decades in upper secondary mathematics classrooms.

In addition to the use of digital technologies in upper secondary (Year 11-12, 16-18-year-olds) mathematics classrooms in Victoria, digital technologies have long been promoted and now expected to be used in the teaching and learning of Year 7-10 mathematics. The rationale of the current mathematics curriculum (VCAA, 2015) notes that digital technologies facilitate the evolution of mathematical ideas “and provide new tools for exploration and invention”. The problem solving proficiency, part of learning in mathematics, explicitly includes developing student capacity to “select and use technological functions” as they “model and investigate problem situations” (VCAA, 2015).

CAS-enabled graphing calculators are mini-mathematical computers with myriad features specifically designed for school mathematics classrooms. More sophisticated CAS-enabled software is used by mathematicians and those whose work involves mathematical modelling (i.e., solving genuine real-world problems). Although the name of the device often focuses attention on the computer algebra system (CAS) capabilities, importantly these devices allow a significant increase in the ways users can interact mathematically. Previously, to explore functions one would need to edit the algebraic representation and investigate the effect on the graphical and numerical representation of the function. With the CAS-enabled technology the user can, for example, 'drag' the graphical representation and explore the effect on the algebraic representation. This dynamic nature and enhanced interactivity provide myriad new opportunities for users (secondary students and teachers) to develop deep mathematical understanding.

In this course, Texas Instruments TI-Nspire CAS CX teacher software was made available to all students free of charge, and used as part of teaching and learning. The teacher software emulates the handheld personal device typically used by secondary students. In addition, one can create screen shots of important ideas to include in coursework learning materials and by sharing the teacher's screen with the software displayed, the teacher can facilitate whole class exploration of important mathematical ideas. In other words, even if the teacher software on the teacher's laptop is all that is available in a particular school classroom, it can still be highly valuable both for the development of mathematical content knowledge (MCK) by the teacher, in the preparation of learning materials, and via projection to enable the whole class to see and exploit the affordances of the technology-rich environment, thus potentially expanding the teacher's pedagogical approach.

Digital technologies for teaching and learning

The second category of digital technologies, that is, *technologies for teaching and learning* more broadly, were also used *by design*. These included cloud-based video conferencing, online notice boards, and interactive whiteboards. These are digital technologies allowing humans to interact synchronously, replicating to some extent the opportunities for teaching and learning offered when a teacher and learners are in the same face-to-face learning space. The face-to-face opportunities include use of concrete materials (including by more than one learner), genuine collaboration (including problem solving, rich access to body language, voice, information), and support. These face-to-face opportunities are also those available in typical school classrooms. In addition, following asynchronous learning, learners can review at least some of the traces of learning (e.g., audio and video recordings, chat), and review, save, and store this for future reference, reflection, and use.

It is expected that in-field teachers of secondary mathematics, in Australia, would have explored various technologies in their university studies preparing them to become teachers of mathematics. In the course designed in this research project, this preparation to teach mathematics with, and value, digital technologies as providing new opportunities for learning was an important consideration in our design. Our intention was to provide a *Technology-Rich Teaching and Learning Environment* (TRTLE) (Brown, 2013), to develop the capacity of OOF teachers to teach mathematics.

Affordances

A Google Scholar search (26 October 2022) confirms that the term 'affordance' is not only commonly used, but this use is increasing. Over 66,000 academic articles use the term. Almost one third of the 'hits' for affordance (i.e., 21 500) are in the last five years (see row 1 in Table 4.1). Row 2 indicates that the articles using the term are increasingly related to education (55% any time to 85% most recently). It is clear from row 3 that a large proportion of these articles are also related to mathematics.

Table 4.1: Search results for affordances over time

Search Term	Anytime	Since 2012 (last 10 Years)	Since 2018 (last 5 years)
Affordance	66 100	33 900	21 500
Affordance + education	39 900	24 300	18 200
Affordance + education + mathematics	33 300	19 000	17 300

However, despite the increasing use—or perhaps because of it—the term affordance is increasingly used without definition or as if all use of the construct is aligned. We use the term following Gibson (1977, 1979) who coined the term. Affordances are what the environment "*offers* the [individual], what it *provides* or *furnishes*, either for good or ill" (Gibson, 1979, p. 119). An affordance is the opportunity for interactivity between *the user* (the actor) and *the object or artefact* for some specific purpose related to learning.

In our study, the object or artefact is digital technology. Following Shulman's (1987) knowledge categories, the learning relates to mathematical content knowledge [MCK] or mathematical pedagogical content knowledge [MPCK] or both. We follow Gibson's linguistic form: verb-[preposition]-

able emphasising the relationship between the actor and the object (e.g., sit-on-able, represent-ability, communicate-ability, reflect-ability). Importantly, to be of use for teaching and learning, affordances must be perceived and enacted (Brown & Stillman, 2014).

Affordance bearers

In considering affordances in digital teaching and learning environments, the specific technology must be considered. Scarantino (2003) uses the term *affordance bearers* to "describe the things in the environment that offer the affordance *in a particular set of circumstances*" (Brown, 2015, p. 113). Affordance bearers can be described in general (e.g., Texas Instruments [TI]-Nspire CAS enabled technology, Zoom). However more specific detail is required "in detailing at a micro level how teachers and students enacted particular affordances" (p. 113), for example, the graphing capability on TI-Nspire or the chat facility in Zoom. The affordance bearer 'Data and Statistics' shown in Figure 4.1 offers the user the affordance *Plot-ability*, only if data has been entered in 'Lists and Spreadsheet' with the variable(s) named. Hence, the affordance is only available in certain circumstances. Further detail and specific examples will be provided later in this chapter.

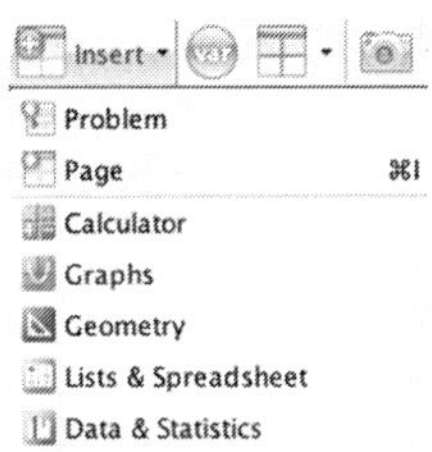

Figure 4.1: Illustrating an affordance bearer

Complexity of design for out-of-field teachers of mathematics

Previous research investigated the perception and enactment of affordances of *Technology-Rich Teaching and Learning Environments* (TRTLEs) for student understanding of function (Brown, 2013). That study included one in-field teacher per mathematics class and clearly specified curriculum content including expectations regarding the use of digital technologies.

The study reported here, in contrast, involves the design of a learning environment to support the learning of OOF teachers with varied:

- Knowledge of mathematical content (i.e., specific content in Years 7-10)
- Mathematical knowledge for teaching (knowledge of common errors and difficulties and how to address these, knowledge of content beyond what is to be taught–breadth and depth)
- Curriculum knowledge, and
- Knowledge of how students learn mathematics.

One challenge was to address the situation described by Maass et al. (2019, p. 303), whereby "approaches promoted by mathematics education researchers differ significantly from the day-to-day practices of teachers". Maass et al. note the persistence in the practice of traditional, or procedural, instruction and argue this stems from many teachers' views of mathematics as being exact, abstract, and unambiguous. These views typically result in an understanding of the role of teacher as being an instructor (imparting knowledge) rather than a facilitator (see Swan, 2005). This gap between the findings from mathematics education research (evidence-based practice related to the teaching and learning of mathematics) and teacher beliefs, and hence their actual practice, remains a substantive challenge for mathematics teacher educators.

Wood's (2002) seminal research exploring mathematics classroom culture generated a framework of the teaching practices and classroom norms required to shift from a 'traditional' classroom environment to one fostering the development of mathematical thinking. She emphasises the interrelated discussion context, role of explainers, role of listeners (teacher and students) and the mathematical thinking in focus. Wood's (2002) framework was used to support teachers to recognise the critical nature of classroom discourse in setting up and implementing an increase in student responsibility as active listener, explainer, and participant. The role of listeners and explainers impacts the level of mathematical thinking.

Following Wood (2002), teaching OOF teachers of mathematics involves engaging them as learners, and additionally enabling them to realise that the approaches to teaching and learning in the university course are the same approaches that should be subsequently used by them in their school classrooms. Teaching and learning mathematics "occurs in an environment where teachers and learners interact, not just with each other but with tasks and digital technologies" (Brown, 2021, p. 3).

Our focus in this chapter is on one of the semester two units in the course focused on 'Knowledge for teaching mathematics 7-10: Statistics and probability'. Taking a design-based approach, our design followed the iterative cycle: design → implement → evaluate → refine (see Collins et al. 2004; Stillman et al., 2009). Thus, as designers, we had the opportunity to complete the first cycle in our design-based research and draw on our learnings from that cycle (Semester 1, 12 weeks) to inform adjustments to the design of the unit. In semester one, the researchers taught in different units, but we came together to reflect on the design (theory and practice) and to teach in this unit together. Furthermore, our students, the OOF teachers, had already immersed themselves in the course and were familiar with the approaches to teaching and learning utilised in the course and subsequently promoted in school classrooms.

Affordances for teaching and learning mathematics

In an earlier study (Brown, 2013) in school classrooms, 16 affordances were identified. One affordance was *Communicate-ability. Communicate-ability* was defined as affordances of a *Technology-Rich Teaching and Learning Environment* (TRTLE) involving support of communication between humans through digital technologies. It is thus a super-category of more specific affordances. This affordance may be initiated by the 'teacher' or the 'student'. The affordance might be for teaching or learning.

As with other affordances[1], *Communicate-ability* was identified in differing circumstances and with differing purposes. It could be manifest via *lesson flow*. This was generally initiated by the 'teacher educator' who might share a page from the unit website for all students to see (affordance for teaching). In addition, it could be manifest via *display of screens,* where either the 'teacher' or 'student' shares the screen of some mathematical technology (e.g., TI-Nspire). Whilst face-to-face, this could be between a pair of students with one literally showing their screen. Online and synchronous, this is either a screen shared with the 'whole class' (the entire cohort of students) or a workshop group of 25 students, or a 'breakout group' of up to 5 students. As TI-Nspire is interactive (versus static), so the sharing could be *interactive*. Display of screens (static) can also be via Padlet where students posted screenshots of important ideas from their mathematical work to illustrate their thinking.

Keeping in mind the mathematical focus on functions in the earlier study (Brown, 2013), two other affordances were:

---

[1] For increased *Read-ability*, affordances are named in title case and italicised.

- *Data Display-ability*: Affordances of TRTLE to provide a graphical display of data (i.e., a plot of numerical data) and manifest in one way, plotting data.
- *Represent-ability:* Affordances of TRTLE involving representations of functions within a representation across representations.

Affordances by design

Regarding affordances for teaching and learning, we were particularly interested in communication, collaboration, and reflections as critical elements of learning. Hence, myriad design elements focused on promoting the perception and enactment of the affordances: *Communicate-ability, Collaborate-ability*, and *Reflect-ability*. The learning time consisted of: preparation, keynote, workshops, breakouts in home groups, and review. Each of these design elements was crucial.

The elaborations and illustrations that follow are intended to give a sense of what the researchers understand by the notion of an affordance in our design. To promote *Reflect-ability*, following a specific learning experience, students were typically asked:

- What were the key mathematics ideas?
- What were the key ideas related to teaching and learning?
- What are the implications for future teaching?

At times, students were expected to reflect in a small group, at other times individually. Reflection occurred post learning activity, and may have been mathematical or pedagogical or a combination of these.

Affordance bearers used for enacting *Reflect-ability* included: cloud-based video conferencing (Zoom), Share Screen, Zoom breakout rooms, digital whiteboards (e.g., Jamboard), online noticeboards (e.g., Padlet), and digital dynamic word clouds (e.g., Mentimeter). The later items in the list could be added to post-class, perused to see what was contributed by others to further learning, and downloaded to add to one's digital notebook of learning in the course. Furthermore, they provided all students with 'a voice'.

Previous work by Brown (2013) focused on identifying affordances. She focused on whether or not the students and teachers perceived particular affordances, and when they did so, if they enacted the affordance or not. In this study, our focus was on designing the course such that key affordances were foregrounded. Our task was to ensure, as best as possible, that particular affordances relevant to the teaching and learning of statistics and probability were 'available' within the learning environment and subsequently perceived and enacted.

One particular affordance for the teaching and learning of statistics is *Identify-ability: Degree of Fit-ability.* This affordance is defined as affordances of TRTLE's to indicate the goodness of mathematical fit of a model to data. Manifestations of this affordance include minimising squares, visual check only, checking with regression coefficient, refining fit by varying parameters of function equation, visual check plus percentage error, and pointwise check. For each manifestation, the affordance bearers used, might vary (e.g., Lists & Spreadsheet, Data & Statistics, Add movable line, Show Residual Squares Window Settings), as will the conditions enabling perception and conditions promoting enactment. Figure 4.2 illustrates some aspects involved in enacting this affordance, in the circumstance of having data entered and plotted.

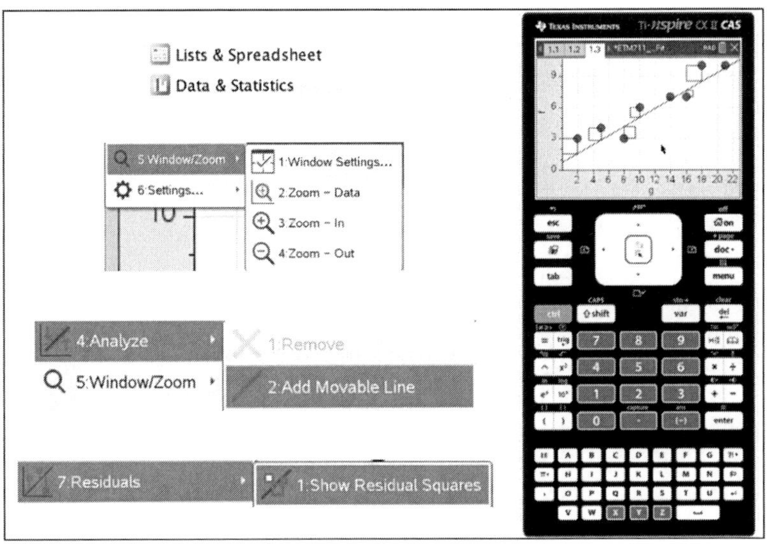

Figure 4.2: Illustrating the affordance *Identify-ability: Degree of Fit-ability*

Figure 4.2 gives a sense of the complexity of technology use, keeping in mind that this parallels the complexity of mathematics itself and hence the teacher MCK required. The opportunities provided for learners to understand concepts is significant. In this example, the statistic measuring the closeness of fit is no longer abstract. Learners can move the line (up, down, or vary the gradient) and consider the impact on the residual squares individually and collectively. Data points can also be 'dragged' and learners hence come to understand how one value can impact the line of good fit. With interactive technologies, there is no reason to focus on procedural knowledge over sense-making and hence conceptual understanding (e.g., Kilpatrick et al., 2001).

The design

Following Liljedahl (2015, p. 629) "as the design team was not primarily a research initiative" the data consists primarily of our field notes and observations, in planning, implementation, and reflection on the implementation. At various phases of each learning activity embedded in the design, the planned focus of intended learning varied. We acknowledge, that just like students in secondary mathematics classrooms, the OOF teachers in the course, that is the learners, can act of their own volition.

In the design, we focused on: tasks, digital technology use, mathematical content knowledge, mathematical pedagogical content knowledge, curricula knowledge, metacognition, recording thinking and more. We identified three phases in the design. There was the whole class focus, the student-student focus, and an independent focus. This is not the structure of the learning activities, that is, these phases overlapped and were woven into the design.

During the student-student focus, collaboration was expected, encouraged and valued as an integral part of learning. The 'teacher' (i.e., one of the mathematics education team who might be the Unit Chair or a Workshop Leader) may participate in this collaboration. This participation might be instigated by the teacher or the students. Given the diversity of experience and knowledge of individual students, different student groups could have quite a different 'focus' during student-student collaboration, within the general learning activity underway. This was expected and planned for in order to best meet the needs of the learners. Moreover, this reflects the approach to differentiation taken in the unit, in which all learners can engage with the same task, acknowledging they will likely end at different places if real learning is to occur. The design ensured all students could access the task, make progress, and learn.

The whole class focus involved a common approach that was taken by all students. This was guided by one or more members of the mathematics education team. This aspect was also designed to ensure important ideas were explored, discussed, and heard by all students.

The independent phase was designed using two approaches. During class, in order to maximise collaboration opportunities, this was generally limited to individual student note-making of thinking (mathematical and pedagogical) and reflecting on their learning. Before each class, students engaged with Keynote recordings, readings, and occasionally attempted mathematics tasks (e.g., a research item from a reading, an assessment item from the national testing program, NAPLAN). Taking a flipped classroom approach (e.g., Stillman, 2017) to the Keynote recordings, allowed students to more fully engage with the content–at a suitable time, and particularly stopping and starting the recording as they considered key ideas before proceeding with the Keynote. With the readings, particular parts were to be read before class with guiding questions. Opportunities were provided in all classes to discuss the Keynotes and readings. After class, students could revisit these (e.g., re-listen, watch, read again, read further, reflect on discussion).

Multiple affordances

In the teaching and learning of mathematics, multiple representations play a critical role as do connections between and within representations. In secondary mathematics, these representations include (but are not limited to) algebraic, graphical, and numerical representations. The learning experience, *Connecting representations: Dot plot and boxplot exploration*, described here include one of the few times a pre-created TI-Nspire file was used. In almost every other situation, students created their own files or used one created by the first author. This was another design feature utilised to maximise learning opportunities.

Figure 4.3 presents two different graphical representations of the same data set. As this is univariate data, a single axis is required, shown here horizontally. It is simple to show the dot plot and the box and whisker plot, one above the other. Dot plots are relatively simple to construct and interpret, in this case each dot represents one height. Box and whisker plots, on the other hand, are complex, particularly to interpret as they are plots of statistics rather than of the data itself (e.g., Watson, 2012).

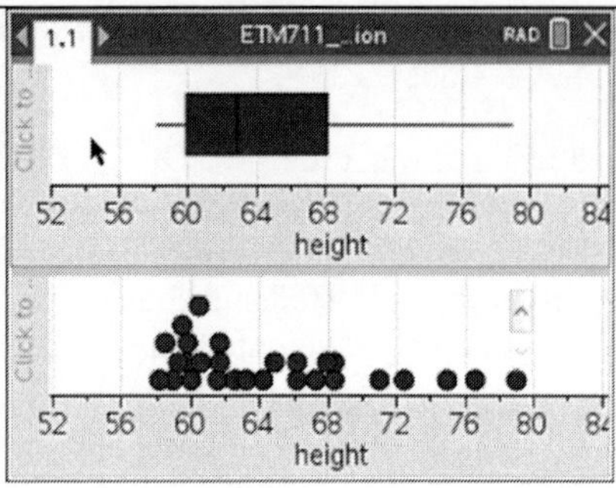	Examine the dot plot and box plot individually and discuss what each tells us. Consider both local [Q1, Q2, Q3, min, max] and global features (quartiles, range, 'shape' - symmetric or skew). How are the plots connected? How would a change in one plot be represented by a change in the other? Explore.
(a) The digital technology	(b) The Task

Figure 4.3: Interactive parallel plots and key elements of the learning activity

During the learning experience, *Connecting representations: Dot plot and Boxplot exploration*, students worked in breakout rooms in small groups. One person was asked to share their TI-Nspire screen to facilitate the exploration and discussion. The task focused on each plot individually and then connections between them. In the dot plot, individual points can be 'dragged'. In the box plot, data in one or more quartiles can be dragged. When one plot is dragged by the user, the other automatically and dynamically updates to represent the changes. The arrows on the right of the dot plot allow the user to reset the plots and data to their original state or undo an action. The reset feature is extremely helpful in looking at relationships between data presented in a dot plot and data summarised in a box plot. Affordances in play in this learning experience include *Communicate-ability, Collaborate-ability, Represent-ability* (Multiple), and *Interact-ability* (dragging point or quartile, effects both plots).

Focusing on the affordances *Communicate-ability* and *Reflect-ability*, Figure 4.4, illustrates a selection of posts to the Padlet by students during or after engagement with this learning activity. The evidence shows a focus both on MCK and MPCK.

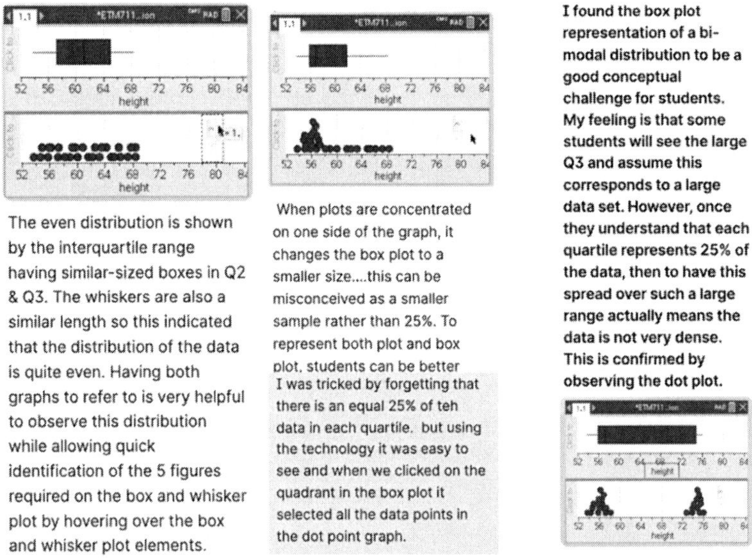

Figure 4.4: Students communicating thinking and reflecting on the task

A word cloud generated on the same day is shown in Figure 4.5. Although the learning experience, *Connecting representations: Dot plot and boxplot exploration* was one of many designed and implemented on that day. The word cloud provides additional evidence of the successful enactment of the affordances, *Communicate-ability* and *Reflect-ability*.

Figure 4.5: Word cloud: Probability and statistics

Concluding remarks

With a focus on methodological approaches, we have presented some of the affordances existing, by design, in the teaching and learning environment provided in a bespoke course for out-of-field teachers of mathematics. If perceived and enacted during the Course and subsequently in school classrooms, these should support the learning of mathematics in school classrooms.

One somewhat unexpected challenge was the beliefs of some of the OOF teachers about their school students. Some, but not all, of this may have been a result of the pandemic, where significant time was spent in lockdown and hence school teaching occurred online. The students, that is the OOF teachers in the course, typically made comments along the lines of:

> My students can't work in groups.
>
> My students are unable to explain their thinking.
>
> The literacy demands [of this learning activity/NAPLAN item] are too high.
>
> The mathematical demands are too high, all/ most of my student are well below level.
>
> My students need to be highly scaffolded.
>
> At my school we take an 'I do, we do, you do' pedagogical approach.
>
> At my school we focus on procedures and [only] when those are learned problem solving.
>
> We chose only fluency and understanding from the [four] proficiencies.
>
> Our students are all on individual learning plans so we can't have the whole class working on the same task.
>
> We follow the textbook/online commercial maths program every/most lessons.
>
> Our students are so low we don't have time for problem solving, discussion, or investigations.

It was important for us that the OOF teachers of mathematics did not apply these views to all their future students as this impacted (i.e., lowered) expectations of learners and reduced approaches to teaching–specifically narrowing the variety of pedagogical approaches used and valued. In our next iterative cycle, teacher beliefs will be a bigger focus as we engage in the refine stage and consider adjustments to the design of the course.

Acknowledgements

We would like to acknowledge Wanty Widjaja and Sandra Herbert, fellow DU mathematics educators who contributed to the overall design of the course, and our team of Workshop Leaders for collaborating with us to implement the design: Tim Alway, Michelle Nicholas, Justine Sakurai, and Jen Waldron.

We wish to thank Texas Instruments Australia who provided the TI-Nspire software to all participants (students and the mathematics education team).

References

Brown, J. P. (2013). *Perceiving and enacting the affordances of Technology-Rich Teaching and Learning Environments (TRTLE's) for student understanding of function* (Unpublished doctoral dissertation). University of Melbourne, Melbourne.

Brown, J. P. (2015). Complexities of digital technology use and the teaching and learning of function. *Computers and Education, 87*, 112-122.

Brown, J. P. (2021). Bridging the gap between evidence-based research, policy, and its implementation in classrooms. In P. White, R. Tytler, J. Ferguson, & J. Cripps Clark (Eds.), *Methodological approaches to STEM education research* (Vol. 2, pp. 1-23). Cambridge Scholars.

Brown, J. P., Sawatzki, C., & Widjaja, W. (2022). Tensions between theory and practice: A mathematics education perspective. In P. White, R. Tytler, J. Ferguson, & J. Cripps Clark (Eds.), *Methodological approaches to STEM education research* (Vol. 3., pp. 286-307). Cambridge Scholars.

Clark-Wilson, A., Robutti, O., & Thomas, M. (2020). Teaching with digital technology. *ZDM – Mathematics Education*, *52*(7), 1223-1242.

Collins, A., Joseph, D., & Bielaczc, K. (2004). Design research: Theoretical and methodological issues. *Journal of the Learning Sciences, 13*(1), 15-42.

Gibson, J. J. (1977). The theory of affordances. In R. Shaw & J. Bransford (Eds,). *Perceiving, acting, and knowing: Toward an ecological psychology* (pp. 67-82). Lawrence Erlbaum.

Gibson, J. J. (1979). *The ecological approach to visual perception.* Houghton Mifflin.

Hobbs, L. (2013). Teaching 'out-of-field' as a boundary crossing event: Factors shaping teacher identity. *International Journal of Science and Mathematics Education, 11*(2), 271-297.

Kilpatrick, J., Swafford, J., & Findell, B. (2001). *Adding it up: Helping children learn mathematics*. National Academy Press. https://nap.nationalacademies.org/read/9822/chapter/1

Leigh-Lancaster, D. (2003). The Victorian Curriculum and Assessment Authority mathematical methods computer algebra pilot study and examinations. In *Proceedings of the 3rd CAME Conference*, (Rheims, France, 23-24 June 2003). https://files.eric.ed.gov/fulltext/ED480463.pdf

Leigh-Lancaster, D., & Stacey, K. (2022). Evolution over two decades of CAS-Active senior secondary mathematics curriculum and assessment. *Mathematics, 10,* 2333.

Liljedahl, P. (2015). Numeracy task design: A case of changing mathematics teaching practice. *ZDM – Mathematics Education*, *47*(4), 625-637.

Maass, K. Cobb, P., Krainer, K., & Potari, D. (2019). Different ways to implement innovative teaching approaches at scale. *Educational Studies in Mathematics, 102*(3), 303-318.

Roschelle, J., Tatar, D., Shechtman, N., & Knudsen, J. (2008). The role of scaling up research in designing for and evaluating robustness. *Educational Studies in Mathematics, 68*(2), 149e170.

Scarantino, A. (2003). Affordances explained. *Philosophy of Science, 70*, 949e961.

Shulman, L. (1987). Knowledge and teaching: Foundations of the new reform. *Harvard Educational Review, 57*(1), 1-22.

Stillman, G. A. (2017). Enabling anticipation through visualisation in mathematising real-world problems in a flipped classroom. In G. Stillman, W. Blum, & G. Kaiser (Eds.), *Mathematical modelling and applications: Crossing and researching boundaries in mathematics education*, (pp. 163-173). Springer.

Stillman, G., Cheung, K., Mason, R., Sheffield, L., Sriraman, B., & Ueno, K. (2009). Challenging mathematics: Classroom practices. In P. Taylor & E. Barbeau (Eds.), *Challenging mathematics in and beyond the classroom* (pp. 243-283). Springer.

Swan, M. (2005). *Improving learning in mathematics: Challenges and strategies*. Teaching and Learning Division, Department for Education and Skills Standards Unit.

Tytler, R., Hobbs, L., Brown, J., White, P., Campbell, C., Cripps Clark, J., Delaney, S., Herbert, S., Peters, A., Sawatzki, C., & Widjaja, W. (2021). Theory and practice relations in design-based research. In P. White, R. Tytler, J. Ferguson, & J. Cripps Clark (Eds.), *Methodological approaches to STEM education research* (Vol. 2. pp. 50-65). Cambridge Scholars.

Victorian Curriculum and Assessment Authority. (2005). *Mathematics Study Design 2006-2015*. VCAA.

Victorian Curriculum and Assessment Authority. (2015). *Mathematics Study Design 2016-2022*. VCAA.

Victorian Curriculum and Assessment Authority. (2016). *The Victorian Curriculum F-10: Mathematics*. VCAA. https://victoriancurriculum.vcaa.vic.edu.au/mathematics/

Watson, J. W. (2012). Box plots in the Australian curriculum. *Australian Mathematics Teacher, 68*(3), 3-11.

Wood, T. (2002). What does it mean to teach mathematics differently? In B. Barton, K. C. Irwin, M. Pfannkuch, & M. Thomas (Eds.), *Mathematics education in the South Pacific* (Proceedings of the 25th annual conference of MERGA, Vol. 1, pp. 61-67), Auckland. Sydney: MERGA.

Jill P. Brown BSc(Ed), GC(Ma&MaEd), MEd, PhD (UniMelb), GCHE (ACU) - Deakin University, School of Education, 221 Burwood Highway, Burwood, 3125, Victoria, Australia.

Jill P. Brown is a mathematics teacher educator and educational researcher in the School of Education. She has more than 20 years' experience in higher education, where she has focused on mathematics and mathematics education for preservice and practising teachers (including out-of-field teachers) in early childhood, primary, and secondary settings. She is the Course Director of Deakin's Graduate Certificate of Secondary Mathematics for out-of-field teachers. Jill's research in the field of mathematical modelling, the teaching and learning of functions, and the use of digital technologies by teachers and students is internationally recognised. She was an elected member on the International Executive of the International Community of Teachers of Mathematical Modelling and Applications (ICTMA) 2013-2022, and is a member of the expert panel for the International Mathematical Modelling Challenge (IMMC), and a member of the Australian IMMC advisory team and judging panel.

https://orcid.org/0000-0002-9202-3634

Carly Sawatzki (BA/BBus(Mgt), GradDipEd(Sec), MEd, PhD) - Deakin University, School of Education, 221 Burwood Highway, Burwood, 3125, Victoria, Australia.

Carly Sawatzki is a teacher educator and educational researcher in the School of Education. She has more than 15 years' experience working with preservice and practising teachers (including out-of-field and non-specialist teachers of mathematics) across primary and secondary courses. Carly's work is distinctly 'real world' and aims to gently influence the way teachers think about educating young people for active and informed citizenship. She is internationally recognised for her classroom research which explores how young people develop numeracy and financial capability within families, communities, and schools. Carly works collaboratively with teachers and students to design interdisciplinary learning tasks situated in contemporary financial contexts. Carly has published in prestigious international journals and led research and curriculum consultancies for Australian, State, and Territory education authorities. She is regularly engaged by teacher associations and schools, being recognised as a thought-provoking presenter who draws on research to challenge thinking, promote critical conversation, and inspire innovation.
https://orcid.org/0000-0002-1463-8844

CHAPTER 5

ENACTING DESIGN-BASED RESEARCH AS A COMMUNITY OF PRACTICE IN PRIMARY SCIENCE: COMMUNITY OF INQUIRY AS A METHODOLOGICAL LENS

MELINDA KIRK & JOSEPH PAUL FERGUSON

Abstract

In this chapter, we demonstrate how university and school research partnerships can move beyond philosophies of boundary crossing, and working at boundaries, towards working and learning together, in order to not only bridge the divide between research and practice but also develop an aligned community of practice. We share insights from enacting a design-based research (DBR) methodology, which was built on reciprocity and a shared direction that focused on improved student learning and efficacy in primary science. Our DBR embodies a shared goal to develop pedagogies that foster students' creativity and critical thinking in science. We adopt a Community of Inquiry lens (CoI) to explore the value of our authentic Community of Practice (CoP) for teachers, researchers, and students alike. We utilise Wenger's (1998) three-part framing of CoP to realise our DBR-enabled community as committed to research partnerships, and we draw on Peirce's (1992, 1998) notion of CoI as a methodological lens to highlight the unique aspects of our DBR as concerning a CoP committed to the realisation of practical impact.

Key words: Design-based research (DBR), Community of Inquiry (CoI), Community of Practice (CoP), Research partnerships

Design-based research

Design-based research is increasingly adopted within the education research community, with the objective to improve science teaching and learning praxis (Juuti & Lavonen, 2006). Many science education researchers advocate for DBR (e.g., Tytler et al., 2021) because it enables rigorous and relevant research (Herrington et al., 2007) which aims to develop both theory and practice (Barab & Squire, 2004). Recent DBR includes (but is not limited to); science learning and technology (Hoadley, 2023), science learning within STEAM (Li et al., 2022), pedagogy in science (Aksela, 2019; Tytler et al., 2022), science teaching-learning sequences (Juuti et al., 2016); co-designed science project-based learning units (Juuti et al., 2021); and university and school partnerships (Coburn et al., 2021; Getenet, 2019). DBR in the science education context consists of three broadly guiding principles. Firstly, DBR is founded on co-participation, enacted by researchers and educators (e.g., school teachers) working together as co-researchers (e.g., Getenet, 2019). Secondly, DBR is flexible and positioned within real-world contexts (e.g., Goff & Getenet, 2017). And thirdly, DBR is iterative and is consistently recognised as a collaborative intervention method (DBIR: Fishman et al., 2013).

The research discussed in this chapter activates these three constructs and is grounded in the socio-cultural perspective that learning and teaching, as well as DBR, are situated and socially supported in practice (Wenger, 1998; Wenger et al., 2002). Our DBR pursues a deepened understanding of a sense of community (Barab & Squire, 2004; Martinovic et al., 2012) as manifesting as an authentic community of practice (CoP) (Wenger, 1998). We make clear the nature of this form of DBR, as potentially a new direction for DBR, by considering it through the lens of Community of Inquiry (CoI) (Burgh & Nichols, 2012; Burgh & Thornton, 2016). We seek to demonstrate how DBR as CoP can activate and support research-practice partnership, a shared inquiry in which key stakeholders come together to collectively engage with an educational challenge, goal, or question.

Our DBR: Context and background

The research explored in this chapter was conducted with approximately 90 Year 6 students, across three classes, from a metropolitan primary school in Australia. Three of the four participating teachers had worked closely with the researchers over the past six years and participated in the Australian Research Council (ARC) project: Interdisciplinary Mathematics and Science

(IMS) Learning project (Tytler et al., 2018-2020). This prior research participation afforded the teachers a deep understanding of the pedagogy of interdisciplinary, guided inquiry, and representational focused learning (Tytler et al., 2022). It is important to note that the teachers had expressed enthusiasm for continuing this research collaboration and requested to participate in further research with this research team.

The research is composed of two independent, but closely aligned, Deakin University research projects: a PhD research project and a School of Education research project. Researchers shared a focus of supporting teachers' pedagogical moves, innovative lesson design features, monitoring and assessment practices which enable the development of critical and creative thinking skills, and collaboration as advocated in current curricula (ACARA, 2022, 2020). Viewed through the lens of CoI, this multi-dimensional stakeholder collaborative research is considered, as we will show, an authentic CoP as it pursues evidence-informed practices through shared intentions and approaches.

A Community of Inquiry (CoI)

In our DBR, we strive to develop a contextual understanding of possibilities to support student creativity, critical thinking, and collaboration within the primary science classroom. We employ the notion of CoI as advocated by Burgh and Nichols (2012) and Burgh and Thornton (2016) for science education, which they developed from Lipman's (1991) CoI pedagogy for philosophy education. Burgh, Nichols and Thornton, as well as Lipman, base their notions of inquiry as a communal process on Charles Peirce's method of science for inquiry (1992, 1998).

The phrase 'community of inquiry' was first coined by Peirce (1992, 1998) to describe inquirers as collaborating, sharing, respecting, and building on each other's ideas, in order to think deeply, critically analyse, and challenge opinions as they generate inferences and justify conclusions (Garrison, 2013). While this framework has been broadly applied in the context of technology and distance learning (Garrison et al., 2014), it has also been adopted in recent science education research focused on meeting diverse student needs (Cheung et al., 2020). While these researchers recognise the interactional relationship of cognitive, social and teaching aspects of a CoI (Garrison et al., 2010) to support student learning, we extend on this in fresh ways to show the value of our approach. Our DBR represents a communal account of inquiry undertaken by researchers, as they collaborate with students and teachers as a Community of Practice. We take seriously Peirce's insistence that inquiry has to be communal to be productive:

> Especially, one man's experience is nothing, if it stands alone. If he sees what others cannot, we call it hallucination. It is not 'my' experience, but 'our' experience that has to be thought of; and this 'us' has indefinite possibilities. (CP 5.402)[1]

Adopting a CoI lens

According to the Peircean (1992, 1998) semiotic pragmatist account of inquiry, we can be successful in our collaborations with teachers and students only if we function as a community. Our research aimed to enable a community that was committed to developing understandings of the knowledge and skills of teaching and learning in primary science. We investigated the focus topics for the lesson sequences (microorganisms and light) in creative and critical ways, whilst developing scientific inquiry and critical and creative thinking with teachers and students. To this end, we implemented the lesson sequences in an iterative way. We were constantly generating, critiquing, refining, and putting into action our joint ideas in a collaborative way. Students' scientifically grounded ideas, and investigation designs, provided a stimulus for collaboratively negotiated, and contextually responsive, decisions by the community. A shared commitment to make critical and creative use of various sign forms (e.g., verbal discussions, written notes, digital diagrams, physical gestures) supported flexibility, and students' actioning of thc method of science. This reflects what Peirce (1992, 1998) referred to as 'the method of science' in that we–the collective–were flexibly and responsively inquiring rather than methodically following a set procedure. Applying a CoI lens to our research seeks to enable insights into how inquiry can stimulate research that is contextually relevant whilst also ensuring reliability of DBR.

A Community of Practice (CoP)

The concept of CoP has socio-cultural origins that are aligned with situated learning theories (Lave, 2004). We adopt the view that knowledge as research is interactionally constructed and supported through rigorous, mutually beneficial, research practice partnerships (Coburn et al., 2021). Shared goals are commonly recognised as significant within DBR (Plomp, 2013) in order to develop a research community (Wenger, 1998). Importantly,

---

[1] CP x.y = *Collected Papers of Charles Sanders Peirce*, volume *x*, paragraph *y*.

authentic sharing is not unidirectional positioning of the researcher as the authority and prescriber of goals, tasks, knowledge or design interventions, with teachers reduced to recipients (Goff & Getenet, 2017; Juuti et al., 2016).

Reciprocity for all research participants and stakeholders is considered the foundation of our CoP approach to DBR. Teachers were co-designers and co-enactors of research-informed practice. Their shared experiences, insights, and knowledge were informative and instrumental for realising rewarding inquiry experiences for students. We value Wenger's (1998) three principles of a coherent community—mutual engagement, joint enterprise, and shared repertoire—as providing both research goals and requirements for the actioning of collective research. As DBR is focused on the process as much as the outcome (Juuti et al., 2016), we position it as the enactment of CoP.

Our CoP framework: Research-Practice Partnerships for DBR

It is significant to note that developing a CoP requires thoughtful and deliberate collaborative development of shared goals over time. In other words, a commitment to, and coalition of, inquiry. In the following section, we outline how Coburn and colleagues' five key features of Research-Practice Partnerships (Coburn et al., 2021) align with a CoP Framework (Wenger, 1998; Wenger et al., 2002), and key attributes of DBR (Penuel et al., 2015; Plomp, 2013) (Table 5.1.). We emphasise that established trust and working relationships are required to enable authentic insights into enacted pedagogy at the classroom level. Intentional research-practice partnership structures were employed to enable the merging of theory and practice (Coburn et al., 2021) within our CoP as DBR. We overlay this analysis of CoP with the CoI lens to make it apparent that a collective commitment to inquiry drives, demonstrates, and authenticates the enactment of productive DBR.

Table 5.1: The key attributes of our DBR (Penuel et al., 2015; Plomp, 2013) aligned with Community of Practice Dimensions (Wenger, 1998) and Coburn and colleagues' (2021) five identified characteristics of Research-Practice Partnerships (RPPs).

Design Based Research (Framing the three commonly identified features e.g., Penuel, et al., 2015; Plomp, 2013)	**Community of Practice Dimensions** (Adopted and adapted from Wenger, 1998)	**Five Features of Research-Practice Partnerships** (Adapted from Coburn et al., 2021)
Co-participation	**MUTUALISTIC** **Building relationships** and establishing roles with others through direct interaction and reflection on previous engagements, interactions, and identities.	**Long term partnership** **Mutualistic** **Intentional strategies** to foster partnerships/relationships
Flexible, situated, and contextual *Conceptualisation, design, implementation*	**JOINT ENTERPRISE** Developing and negotiating group activities, including reconciling different interpretations, and holding each other to account.	**Focus on addressing challenges** *and developing practices*
Iterative *Collaborative revision and adaptations and re-enacting*	**SHARED REPERTOIRE** Developing shared understandings, shared "routines, words, tools, ways of doing things, stories, gestures, symbols, genres, actions, or concepts" (Wenger, 1998, p. 83)	**Outcome**: Original *(also ongoing)* analysis

Our DBR–Enablers and blockers of CoP

In the following section, we expand on the framing in the above table (Table 5.1) with the understanding that the principles of Research-Practice Partnerships (Coburn et al., 2021), principles of a CoP and the key features of DBR, are not sequential or a tick list. Rather, they are informed by the shared inquiry and are recursive in nature. There is a growing body of literature focused on research-practice partnerships that explores the ways in which partnerships between schools and universities facilitate education research and the professional development of teachers as co-researchers (Beveridge et al., 2018; Martinovic et al., 2012). We seek to make explicit what is brought to light by viewing this DBR/CoP relationship through the lens of CoI.

Mutualistic: Developing research alignment

To conduct research in a primary school is a privilege that requires trust from the school leadership, staff, students, and school community. As such, it is not only important to have positive and open communication with teacher participants but all stakeholders. All project researchers had previously researched for many years at the school that is the focus of this chapter and established working relationships with school leadership and teachers. We recognise that such positive relationships require continual fostering and commitment, they can never be assumed. We argue that the importance of such relationships for our DBR is revealed through the lens of CoI, which makes evident the necessity for CoP to prioritise evidence-informed inquiry as leading to desired changes in the habits of teaching and learning.

Aligning policy, curriculum, school, and teacher goals

We acknowledge that the consideration of CoP stakeholder needs is imperative to establish not only trust, but also reciprocity at the school community level. The focus of the reported research–critical and creative thinking within an interdisciplinary science and mathematics-guided inquiry learning design–emerged from the previous IMS project (Tytler et al., 2018-2020). The development of resources, as well as assessment and reporting methods, was a mutually identified objective. Teachers and researchers recognised a need for curriculum innovation and advocacy for inquiry science along with the general capabilities of critical and creative thinking (ACARA, 2020). Our DBR responds to teachers' feedback and requests to enable, monitor, and assess student critical and creative thinking within primary science. Collectively, we aimed to build on established relationships, and we valued teacher agency as the key driver of authentic CoP as a collective commitment to research as inquiry-informed practice.

The pursuit of school and research alignment is conducive to establishing a good fit for conducting productive research (Beveridge et al., 2018). Flexibility is essential to enable responsiveness to emergent contextual demands of such school-based research (Design-Based Research Collective, 2003; Herrington et al., 2007). Research project alignment with education policy and school identified improvement goals, as part of the School Strategic Plan (e.g., Australia, Victoria Education), was particularly important for our DBR. This was evidenced by school leadership (school principal and leadership team) reporting outcomes of our DBR as evidence of meeting explicitly identified and documented School Strategic Plan and Annual Implementation Plan (DETV, 2021) goals, as part of their formal panel assessed school review. This level of contextually responsive research alignment notably fosters both school and teacher 'buy-in', which is conducive to developing a mutually beneficial DBR for all stakeholders.

Education research broadly recognises the significance of positive university and school relationships (Oates & Bignell, 2022). School leaders' and teachers' agency and continued co-participation in research are made possible by attention to evidence, including that of previous research projects. Longitudinal research-practice partnerships committed to shared inquiry were evidenced by school leaders' and teachers' requests to maintain our research partnership and continue to co-research beyond the research reported in this chapter. Considerations for facilitating research alignment as part of the development of a shared inquiry are unpacked in the following section.

Joint enterprise: Intentional and flexible practices

Through the lens of CoI, we can see that a joint commitment to inquiry as part of the enterprise of enacting the method of science, is integral to establishing a shared enterprise. Before commencing a new project, even within established partnerships, it was essential to collaboratively develop both collective alignment and direction. The planning outline below (Table 5.2) identifies considerations and processes that were viewed as essential for setting up our collective research partnership.

The following questions (refer to Table 5.2) were discussed with participating stakeholders before research commenced–in order to foster a joint enterprise–and throughout the research journey for purposes of collaborative reflection.

Table 5.2: Initial DBR planning outline to determine our orientation for a shared direction

Developing research alignment	Developing shared and agreed structures
Why? Purpose, clarification, and alignment.	**Sharing and Unpacking:** What are stakeholders' participation interests and goals? How could the research support school identify improvement/ goals? And school community? What are all participants', teacher-researcher goals for participation? How would they like to be supported? Questions and concerns?
What? Curriculum and scientific concepts with attention to critical and creative thinking and the role of collaboration (e.g., collaborative inquiry) (the focus of the research project).	**Previous iterations:** What would you (the teachers) like to be included/not included/changed? **Curriculum:** Interdisciplinary opportunities? Where does critical and creative thinking fit in? Where are the opportunities for collaboration?
How? Organisation, pedagogy, and reporting (including monitoring, supporting, and assessing).	**Organisational:** Planning and reflection workshops (3) and weekly (12) meetings and communications to ascertain: timeline, timetables, resources, and lesson plan adjustment and review.

	Contextual considerations–including student needs and safety (e.g., coming out of COVID restrictions, student needs, staff, and student absences). **Pedagogical:** Ways of teaching, collaborative inquiry, representational, guided inquiry (IMS approach iteration). **Monitoring and Review:** Ways of monitoring and assessing–student representations and inquiry processes, teacher anecdotal records and informal reflection discussions pre/post lessons (12 weeks). Other correspondence prior to and following lessons as needed (teacher/researcher instigated email/phone/text). **Project reporting:** Interviews (students, teachers), final teacher workshop, school leadership project reflection discussions, possible co-dissemination.

Research collaboration and communication

Both the 'joint enterprise' and 'shared repertoire' within our CoP (Wenger, 1998, 2000, 2002) required monitoring and flexibility that was enabled by communication. Effective and open communication throughout the entire research project was necessary for promoting and upholding our CoP as DBR. Collaborative conversations enabled authentic co-research at all stages: the inquiry learning design, pedagogical actioning, reflection, and revision processes. Inquiry as a driver of DBR as CoP was made possible by purposeful and respectful collaboration between teachers and researchers that facilitated the enactment of the method of science as underpinned by reflective practice (Schön, 2017). Reflection discussions after each lesson provided insights into teacher-identified issues:

a) effective lesson attributes or strategies to be continued/included,
b) student and/or lesson needs and considerations for future lessons,
c) suggestions/adjustment for future lessons and teaching,
d) concerns or challenges to be discussed.

It is important to note that while these were often researcher-instigated conversations, teachers also initiated conversations which were significant for realising our DBR as a shared enterprise. These conversations were possible because the researchers made themselves available to teachers as co-present in the classrooms and were very explicit about valuing the teachers' input. This included in-situ face-to-face conversations initiated by both teachers and/or researchers before, during, and after lessons as needed. Conversations also regularly took place via phone calls, texts, email, or Zoom meetings. Through these open and multiple modes of communication, we were able to collaboratively respond to contextual needs and build our CoP.

Within our CoP, as framed in-situ, ideas were collaboratively revised and adopted in an iterative process. An example of this is when one of the teachers suggested a ‘party’ or celebration as part of the culminating lesson of the microorganism sequence. This idea was discussed and then collaboratively developed to culminate in a ‘Scientists for solutions: Celebratory microorganisms morning tea’. Collaboratively building on the initial party idea, a microbial (learning focus) derived menu was co-developed by teachers and researchers, including yoghurt and cheese (bacteria cultures), hot bread (yeast–fungi), and Gurbuchi (Indigenous Australian Kombucha like drink–yeast and bacteria). Students, teachers, and researchers were excited by the potential of this event; it was a highly anticipated culmination to the sequence that enabled students to experience positive first-hand applications of microorganisms in the real world. The inquiry journey as built on a commitment to the method of science as a path to practical findings is not predetermined, but emergent, as the CoI lens make clear.

Shared repertoire: Extending our CoP

Our research embraced the perspective that DBR is not based on solving specific problems, but rather the appreciation that outcomes can be unexpected (Goff & Getenet, 2017) and emerge as a result of acute attention to evidence and context. Significantly, collaborative deliberation and co-design (Severance et al., 2016) was required to consider how to proceed with an inquiry-based microorganisms learning sequence during the challenging context of the COVID pandemic. Students and teachers were wearing masks, air purifiers were being installed in all classrooms, epidemiologists were regularly reporting on the pandemic in the media, and teachers and students were regularly absent due to illness. The previous pre-pandemic microorganism sequence iteration included students' microbial investigation of surfaces throughout the school. Although teacher-researcher participants had requested for students to be able to conduct an agar plate and microbial inquiry again through surface exposure, this required revision as participant safety was a concern for researchers. Student safety and mental health considerations were paramount, as the students had only recently resumed attending school after intermittently attending for two years (2020-2021) during the pandemic.

Expert contemporary scientist

The COVID context resulted in the research team extending its membership to include a well-known Deakin University epidemiologist, who collaborated with researchers to record a presentation for the students. Students and teachers were familiar with this epidemiologist as she regularly appeared in the media in regard to the pandemic. The addition of an epidemiologist to the community was a contextual response, aligned with the previous microorganism sequence iteration that included a historical scientific perspective of microorganisms through the example of Doctor Joseph Lister (1827-1912). As viewed through the lens of CoI, the inclusion of a contemporary science expert–epidemiologist–boosted the scientific knowledge and skills that define the method of science as the best path for inquiry. This addition provided a powerful, and contextually responsive way, to develop the inquiry practices of our CoP.

Evidence of a student-centred CoP and CoI lens

The adoption of an 'epidemiologist perspective' resulted in an unexpected learning opportunity in the form of students' agentic pursuit of solutions. The students' investigations identified high microbial load in the air of the school lost property boxes, which stimulated their taking on the role of epidemiologists to minimise risks for the school community. Through this inquiry, students realised meaningful science agency; they enacted inquiry that mirrored the inquiry of the researchers and teachers. Through our CoI lens, we can say that inquiry begat inquiry within our COP. This positive outcome resulted in the modification of the learning sequence to further support and value student voice, which was a school-identified goal (to be reported in other publications). The students' evidence-informed proposal for changes to lost property box procedures was adopted by the broader school community. The school leadership reported this experience to be transformative for students, teachers and researchers, formally reporting in the School Annual Review (report) these transformations as evidence of student voice. This adaptive response to demands of context was enabled by collaborative partnerships formed with the broader Deakin University science community, with the epidemiologist further driving the need for the CoP to prioritise inquiry as the method of science.

Shared identities

The reciprocal nature of our research-practice partnership resulted in both the teachers identifying as co-researchers, and significantly the researchers identifying as co-teachers. We propose, with our CoI lens again at work, that this process of identity formation was made possible by a commitment to inquiry as teachers and researchers alike were enacting the method of science to inform changes in practice. This sharing of teacher identity was initiated by the case study teacher, as she invited the researchers to co-teach as part of a teaching-team approach. The research team were considered by teachers as: 'teachers researching collaboratively with the class', and 'researchers from the university collaborating and learning together with the class'. Similarly, researchers viewed both teachers and students as co-researchers–sharing in a commitment to the research and the collaborative inquiry ethos. In our CoP, this entailed the exploration of teachers' ideas prior to lessons as well as considering and adopting student inquiry ideas as appropriate. Dialogic interactions as part of this process included insitu-reflective discussions between the teacher, researchers and students, as initiated by the teacher. In response to student ideas, DBR methods were actioned that reflected a synergy of research and teaching. Both researchers and teachers exhibited trust and authentic sharing within the classroom environment. This was pivotal to the enactment of our CoP and indicative that we (researchers) were not simply viewed as visiting outsiders (Coburn et al., 2008) by teachers and students, rather we were positioned as education partners on an inquiry journey.

Analysis and research reliability

We contend that these overlapping roles of co-researcher and co-teacher enabled a closeness to the data. We were all committed to inquiry as the method of science that amplified teacher perspectives, pedagogical enactment, and foregrounded contextual applicability. Although some research suggests that such closeness to data risks bias and validity (e.g., Dede, 2004), we posit that DBR participant shared identities and roles, can enable a closeness to the data that adds legitimacy rather than compromises the validity of the research. A CoP as DBR approach as framed by CoI can provide an opportunity for what we term 'thick data' and 'thick reporting' or research narration (Bell et al., 2004; Juuti & Lavonen, 2006). Our immersion in the school and classroom context with teachers and students served to ensure relevancy and reliability of the findings. However, methodological awareness was required to avoid slipping into bias. The CoI lens makes clear that the standards of the method of science must be upheld in order for inquiry to maintain its epistemic integrity.

With a focus on review and revision, DBR as narration provided a juncture point for design and pedagogical moves (Hoadley & Campos, 2022). Teacher willingness to openly share their insights was instrumental to refining practice in productive ways as informed by research as inquiry. Open and collaborative relationships with our research participants minimised the risk of demagogic responses to the research, which would otherwise threaten the integrity of the inquiry that underpins our CoP as DBR. This was evidenced when the teachers were asked about the inclusion of the pedagogical language in the first draft lesson plan. Teachers responded that although they agreed with the pedagogical stages (Tytler et al., 2022), their inclusion within each lesson plan 'wasn't adding anything, other than making the document laborious'. Consequently, we substituted the IMS specific pedagogy sectional language with an agreed colour-coding scheme, to represent the IMS pedagogy stages. As such, we recognised the pedagogical alignment but ensured ease of lesson plan reading. This is indicative of the authentic and intimate knowledge that characterises the co-researcher identity (Hoadley, 2004; Hoadley & Compos, 2022), made evident with the CoI lens that emphasises the alignment of rationale and practice (Hoadley, 2004; Hoadley & Campos, 2022).

Conclusion

This research illustrates, through the methodological lens of CoI, the way in which CoP as a particular enactment of DBR enables informed and authentic adaptation of the design, implementation, and analysis of inquiry science. Alignment with school goals, policy and curriculum demands was realised through collaborative contextual responsiveness. In particular, this included enabling student inquiry as aligned with the inquiry of the researchers and teachers. A shared commitment to inquiry–the method of science–was adopted in order to change habits of teaching and learning that are fundamental to our contextually responsive enactment of DBR.

A potentially transformative CoP emerged as a result of the authentic sharing of identities and consideration of school and classroom context within the broader pandemic context. The sharing of both teacher and researcher roles was made possible by long established and positive relationships between the school and university, teachers, and researchers. All involved were driven to employ the method of science to realise inquiry as a way to make proactive changes. The closeness with the research that this approach afforded, illustrates the methodological power and possibilities of DBR as a CoP.

It is significant to note that this form of DBR is a longitudinal pursuit. Collaborative partnerships cannot be assumed, rather they require deliberate facilitation and continued maintenance over time. To enable pro-active and productive longitudinal research-practice partnerships, we advocate for school-based research to adopt a flexible CoP framing. Within our DBR, this approach prioritises research as shared inquiry through the method of science. Although this chapter has not provided a micro-analysis of our DBR processes or outcomes, we hope our insights can inspire more school-based research to adopt a CoP approach with a CoI methodological lens. In contrast to an educationally detached conception of boundaries, intersections, and bridges (to be crossed) (e.g., Akkerman & Bakker, 2011; Farrell et al., 2022), or a 'them and us' framing, our COP as DBR promotes a research coalition. It is this rationale that can establish research relevance, reliability and the possibility of transformative outcomes. We propose that a CoI methodological lens is highly valuable to stimulate, support, and foster DBR as a means to realise impact, to generate practical and evidence-informed changes.

Acknowledgements

We thank the participating school and the Victorian Department of Education for making this research possible. A special acknowledgement and thanks to Professor Catherine Bennett for her contribution.

References

Akkerman, S. F., & Bakker, A. (2011). Boundary crossing and boundary objects. *Review of Educational Research, 81*(2), 132-169.

Aksela, M. (2019). Towards student-centred solutions and pedagogical innovations in science education through co-design approach within design-based research. *LUMAT: International Journal on Math, Science and Technology Education*, *7*(3), 113-139.

Australian Curriculum, Assessment and Reporting Authority. [ACARA]. (2020). Curriculum review process paper: Version 1.0. ACARA: Sydney. https://www.acara.edu.au/docs/default-source/curriculum/curriculum-review-process-paper-31-august-2020.pdf

Australian Curriculum and Assessment Reporting Authority. [ACARA]. (2022). The Australian curriculum: Science foundation to Year 10. https://www.australiancurriculum.edu.au/f-10- curriculum/science

Barab, S., & Squire, K. (2004). Design-based research: Putting a stake in the ground. *The Journal of the Learning Sciences, 13*(1), 1-14.

Bell, P. (2004). On the Theoretical Breadth of Design-Based Research in Education. *Educational Psychologist, 39*(4), 243-253.

Beveridge, L., Mockler, N., & Gore, J. (2018). An Australian view of the academic partner role in schools, *Educational Action Research, 26*(1), 25-41.

Burgh, G., & Nichols, K. (2012), The parallels between philosophical inquiry and scientific inquiry: Implications for science education. *Educational Philosophy and Theory, 44*(10), 1045-1059.

Burgh, G., & Thornton, S. (2016). Inoculation against wonder: Finding an antidote in Camus, pragmatism and the community of inquiry. *Educational Philosophy and Theory, 48*(9), 884-898.

Cheung, D. H.-C., Ng, A. K.-L., Kiang, K.-M., & Chan, H. H.-Y. (2020). Creating a community of inquiry in the science classroom: An effective pedagogy for teaching diverse students? *Journal of Further and Higher Education, 44*(1), 1-13.

Coburn, C. E., Bae, S., & Turner, E. O. (2008). Authority, status, and the dynamics of insider-outsider partnerships at the district level. *Peabody Journal of Education*, *83*, 364-399.

Coburn, C. E., Penuel, W. R., & Farrell, C. C. (2021). Fostering educational improvement with research-practice partnerships. *Phi Delta Kappan, 102*(7), 14-19.

Dede, C. (2004). If design-based research is the answer, what is the question? A commentary on Collins, Joseph, and Bielaczyc; diSessa and

Cobb; and Fishman, Marx, Blumenthal, Kraj- cik, and soloway in the JLS special issue on design-based research. *The Journal of Learning Sciences, 13,* 105-115.

Department of Education and Training Victoria. [DETV]. (2021). School Strategic Plan, Victoria, Australia. https://www2.education.vic.gov.au/pal/school-strategic-plan

Design-Based Research Collective. (2003). Design-Based Research: An Emerging Paradigm for Educational Inquiry. *Educational Researcher, 32*(1), 5-8.

Farrell, C. C., Penuel, W. R., Allen, A., Anderson, E. R., Bohannon, A. X., Coburn, C. E., & Brown, S. L. (2022). Learning at the boundaries of research and practice: A framework for understanding research–practice partnerships. *Educational Researcher* (Advance online publication).

Fishman, B. J., Penuel, W. R., Allen, A. R., Cheng, B. H., & Sabelli, N. O. R. A. (2013). Design-based implementation research: An emerging model for transforming the relationship of research and practice. *Teachers College Record, 115*(14), 136-156.

Garrison, D. R. (2013). Theoretical foundations and epistemological insights of the community of inquiry. *Educational communities of inquiry: Theoretical framework, research and practice* (pp. 1-11). IGI Global.

Garrison, D. R., Anderson, T., & Walter, A. (2010). The first decade of the community of inquiry framework: A retrospective. *Internet and Higher Education, 13* (1-2), 5-9.

Garrison, D. R., Cleveland-Innes, M., & Vaughan, N. (2014). Community of inquiry. *The SAGE encyclopedia of action research, 1*, 147-150.

Getenet, S. (2019). Using design-based research to bring partnership between researchers and practitioners. *Educational Research, 61*(4), 482-494.

Goff, W., & Getenet, S. T. (2017). Design based research in doctoral studies: Adding a new dimension to doctoral research. *International Journal of Doctoral Studies, 12*, 107-121.

Herrington, J., McKenney, S., Reeves, T., & Oliver, R. (2007, June). Design-based research and doctoral students: Guidelines for preparing a dissertation proposal. In *EdMedia+ Innovate Learning* (pp. 4089-4097). Association for the Advancement of Computing in Education (AACE).

Hoadley, C. M. (2004). Methodological alignment in design-based research. *Educational Psychologist, 39,* 203-212.

Hoadley, C. P. (2023). Creating context: Design-based research in creating and understanding CSCL. *Computer support for collaborative learning* (pp. 453-462). Routledge.

Hoadley, C., & Campos, F. C. (2022). Design-based research: What it is and why it matters to studying online learning, *Educational Psychologist, 57*(3), 207-220.

Juuti, K., & Lavonen, J. (2006). Design-based research in science education: One step towards methodology. *Nordina, 4*, 54-68.

Juuti, K., Lavonen, J., & Meisalo, V. (2016). Pragmatic design-based research – Designing as a shared activity of teachers and researchers. In D. Psillos, & P. Kariotoglou (Eds.), *Iterative design of teaching-learning sequences*. Springer.

Juuti, K., Lavonen, J., Salonen, V., Salmela-Aro, K., Schneider, B., & Krajcik, J. (2021). A teacher–researcher partnership for professional learning: Co-designing project-based learning units to increase student engagement in science classes. *Journal of Science Teacher Education, 32*(6), 625-641.

Lave, J. (2004). Situating learning in communities of practice. In L. Resnick, J. Levine, & S. Teasley (Eds.), *Perspectives on socially shared cognition* (pp. 63-84). American Psychological Association.

Li, J., Luo, H., Zhao, L., Zhu, M., Ma, L., & Liao, X. (2022). Promoting STEAM education in primary school through cooperative teaching: A design-based research study. *Sustainability 14*, 10333.

Lipman, M. (1991). *Thinking in education*. Cambridge University Press.

Martinovic, D., Wiebe, N., Ratkovic, S., Willard-Holt, C., Spencer, T., & Cantalini-Williams, M. (2012). 'Doing research was inspiring': Building a research community with teachers. *Educational Action Research, 20*(3), 385-406.

Oates, C., & Bignell, C. (2022). School and university in partnership: A shared enquiry into teachers' collaborative practices, *Professional Development in Education, (48)*1, 105-119.

Penuel, W. R., Allen, A. R., Coburn, C. E., & Farrell, C. (2015). Conceptualizing research–Practice partnerships as joint work at boundaries. *Journal of Education for Students Placed at Risk (JESPAR), 20*(1-2), 182-197.

Peirce, C. S. (1992). *The essential Peirce — Selected philosophical writings (1867–1893) (Vol. 1)*, N. Houser & C. Kloesel (Eds.). Indiana University Press.

Peirce, C. S. (1998). *The essential Peirce — Selected philosophical writings (1893– 1913) (Vol. 2)*, N. Houser, A. De Tienne, J. R. Eller, C. L. Clark, A. C. Lewis, & D. B. Davis (Eds.). Indiana University Press.

Plomp, T. (2013). Educational design research: An introduction. In T. Plomp, & N. Nieveen (Eds.), *An introduction to educational design research* (pp. 9-36). Netherlands Institute for Curriculum Development (SLO).

Schön, D. A. (2017). *The reflective practitioner: How professionals think in action*. Routledge.

Severance, S., Penuel, W. R., Sumner, T., & Leary, H. (2016). Organizing for teacher agency in curricular co-design. *Journal of the Learning Sciences, 25*(4), 531-564.

Tytler, R., Hobbs, L., Brown, J., White, P. J., Campbell, C., Cripps Clark, J., Delaney, S., Herbert, S., Peters, A., Sawatzki, C., & Widjaja, W. (2021). Theory and practice relations in design-based research: Designing professional learning with teachers teaching mathematics and science out-of-field. In P. White, R. Tytler, J. P. Ferguson, & J. Cripps Clark (Eds.), *Methodological approaches to STEM education research – Volume 2* (pp. 40-65). Cambridge Scholars Publishing.

Tytler, R., Mulligan, J., Prain, V., White, P., Xu, L, Kirk, M., Nielsen, C., & Speldewinde, C. (2021). An interdisciplinary approach to primary school mathematics and science learning. *International Journal of Science Education*, *43*(12), 1926-1949.

Tytler, R., Prain, V., Kirk, M., Mulligan, J., Nielsen, C., Speldewinde, C., White, P., Xu, L. (2022). Characterising a representation construction pedagogy for integrating science and mathematics in the primary school. *International Journal of Science and Mathematics Education.*

Tytler, R., White, P., Xu, L., Prain, V., Mulligan, J., Lehrer, R., Schauble, L., Kirk, M, Neilsen, C., & Speldewinde, C. (2018-2020*). Interdisciplinary mathematics and science learning project.* https://imslearning.org

Wenger, E. (1998). *Communities of practice. Learning, meaning and identity*. Cambridge University Press.

Wenger, E., McDermott, R. A., & Snyder, W. (2002). *Cultivating communities of practice: A guide to managing knowledge.* Harvard Business School Press.

Wenger, E., & Snyder, W. (2000). Communities of practice: The organizational frontier. *Harvard Business Review,* Jan-Feb, 139-45.

Melinda Kirk (BEd, MEd) - Deakin University, School of Education, Faculty of Arts & Education, 221 Burwood Highway, Burwood, 3125, Victoria, Australia.

Melinda Kirk is a PhD candidate, educational researcher, and sessional lecturer. An experienced educator she has a special interest in primary interdisciplinary and creative education. Her PhD research focuses on the

way in which critical and creative thinking can be enabled, monitored, and supported in the primary interdisciplinary classroom. Melinda is particularly interested in interdisciplinarity and enabling student voice and agency through co-developing lesson design and pedagogy with teachers and students.
https://orcid.org/0000-0002-0058-6289

Joseph Paul Ferguson (BSc, BA(Hons), MTeach, PhD) - Deakin University, School of Education, Faculty of Arts & Education, 221 Burwood Highway, Burwood, 3125, Victoria, Australia.

Joseph Paul Ferguson is an educational researcher and teacher educator interested in investigating the various manifestations of reasoning in the science classroom, in particular creative reasoning. Through an exploration of the philosophies and theories that underpin inquiry practices and the associated methodologies that enable this research, Joseph seeks to add to efforts to clarify what it means for teachers and students to enact inquiry in the classroom. Joseph is particularly interested in the affordances of video-based methodologies, informed by film theory and philosophy, to maximise the potential of such research to support teachers and students in their teaching and learning of science.
https://orcid.org/0000-0002-0971-3256

CHAPTER 6

A PROMETHEUS OF OUR TIMES: INVESTIGATING THE SCIENCE CLASSROOM THROUGH PEIRCE, FRANKLIN, AND FILM

JOSEPH PAUL FERGUSON
& JOHN CRIPPS CLARK

Abstract

The extensive work of Charles Peirce has failed to gain significant traction in science education research. This is despite Peirce's status as a practising scientist who philosophised with the method of science in mind. Peirce offers us deep insights into science education, but due to the perceived gap between his pragmatist semiotics and the 'messy' realities of the classroom this has yet to be realised in ways that enrich our methodological work as we struggle to realise new perspectives on old problems. We have looked to film for methodological inspiration, and recently viewed Ken Burns' documentary 'Benjamin Franklin' in which Franklin, for us, provides an exemplar through which to interrogate Peirce's philosophy for insight into our research methodology as a semiotic process of diverse sign forms. We argue that Franklin personifies, in a sometimes-paradoxical way, Peirce's account of truth as the ultimate aim of inquiry that is realised as/through signs. We explore the methodological implications of taking seriously Peirce's correspondence, coherence, consensus, and instrumentalist notions of truth for our research and teaching, as well as claiming that film ought to be part of our methodological toolkit.

Key words: Film semiotics, Pragmatism, Peirce, Scientific truth, Science education

Film as methodological refraction

In our work as researchers and teachers of science education, we are constantly seeking inspiration to productively engage with the methodological challenges that arise as we strive to research and teach in ways that make a theoretical and practical difference. Therefore, not surprisingly, as cinephiles we often look to films for methodological insights (Cripps Clark & Ferguson, 2020) as we strive to expand traditional methodological approaches, which are predominately textual in nature, with the semiotic richness and diversity of films as constellations of audio-visual signs. This is film not only as recreation or escape, but film also as visual and auditory provocation that enables us to think, see, hear and feel in different ways. In other words, film as methodological refraction. This refraction of film reveals the social and technological forces that have shaped it and us, and what is revealed and hidden by film then stimulates us to investigate not just the presences but also the absences in our understandings. Openness to the refractive power of film is valuable for us in our endeavours to grapple with the convoluted philosophical groundings of our research and teaching, none more so than the works of Charles Peirce (Ferguson & Cripps Clark, 2021) which are notoriously difficult to understand in the written form and hence the need for engaging with his ideas through an array of representational forms (including film). In this capacity, Ken Burns' documentary 'Benjamin Franklin' enabled us to approach anew Peirce's complex account of truth as the ultimate aim of inquiry that is realised as/through signs, which has potentially significant ramifications for our methodological practices as science education researchers and teachers. However, as we will show, Burns' filmic version of Franklin is not complete in the sense that to fully appreciate Franklin as the personification of Peirce's notion of truth requires us to engage with myriad representations of Franklin that vary in form and content.

Before proceeding, it is necessary to offer some words about Burns' approach to filmmaking and the way in which his films brim with epistemic potential and as such are a worthy tool in our methodological endeavours. Burns' documentaries (Thelan, 1994) are thoroughly researched and intentionally educative in nature–financially supported by publicly minded institutions–as he seeks to inform the public about the important people and events that constitute American history. As such, he values his films not so much in terms of the revenue they generate but rather the degree to which they impact the American psyche, and we should treat them in the same way. We thus suggest that Burns' films can be viewed as academic artefacts, or public service announcements, much like those that we generate as part of our research. And just like our academic papers and presentations, Burns adopts a particular way of constructing filmic arguments that serves his agenda of instilling Americans with a critical respect towards the people and events in the history of the U.S.A. This is an approach that fixates on individuals, such as Benjamin Franklin, as tragic, heroic, and transcendental figures who ‘made America great’. Burns’ modus operandi is to mythologise individual Americans of the past, and he does so by ‘activating’ still photographs and similar static imagery (e.g., paintings, engravings) through a dynamic process of slowly zooming in on subjects of interest and panning from one subject to another. By using this so called ‘Ken Burns effect’, he is able to ‘bring to life’ America’s historical figures, many of whom preceded videography and photography by hundreds of years. The narrators’ voices chosen to accompany these images (e.g., Peter Coyote, Tom Hanks, Whoopi Goldberg) are always full of authority, pride, sadness, hospitality, and most importantly are ‘authentically’ American. In watching and listening to a Ken Burns film, you experience key thinkers and doers as well as their revolutionary ideas in new ways, which are productive of an ongoing consideration of how we ought to undertake historically, theoretically, and philosophically informed research.

Peirce’s understanding of truth and reality

Charles Peirce was a founder of pragmatism, which over his lifetime as a scientist and philosopher he developed into what has become the third great philosophical tradition (Legg & Hookaway, 2021). Peirce’s philosophy, in the tradition of Immanuel Kant, was all-encompassing, but fundamentally his aim was teleological; to determine the nature and value of science as a method of inquiry that enables the establishment of the truth.

> The opinion which is fated to be ultimately agreed to by all who investigate, is what we mean by the truth, and the object represented in this opinion is the real. (CP 5.407)[1]

Thus, as Lane (2018) points out, Peirce’s notions of *truth* and *reality* are distinct but mutually constitutive, but it is his account of truth that is most relevant for our interest in science education because it is truth that is directly tied to inquiry through/as the creation and interpretation of signs.

---

[1] In referencing Peirce’s work in this chapter, we use his *Collected Papers* whenever possible. In doing so, we use the standard protocol: CP x.y = *Collected Papers of Charles Sanders Peirce*, volume *x*. paragraph *y*.

For the purposes of this chapter, all we need to know about Peirce's notion of reality is:

> It is important to observe that the essential difference between a reality and nonreality, is that the former has an existence entirely independent of what you or I or any number of men may think about it. (W2: 104)[2]

Lane (2018) calls Peirce's account of truth, a "dual-aspect account of truth" (p. 13), concerning as it does the semiotic (signs of meaning) and pragmatist (inquiry as habits of action based on the effects of the problem at-hand) nature of truth. In this way, truth has a "representationalist aspect" and an "investigative aspect" (p. 8). As Forster (2011) makes clear, for Peirce "the question of the nature of true symbols is inseparable from questions of meaning and method addressed in the science of inquiry" (p. 160). It is symbols as propositions, among the three symbol types (the others are icons and indices), which are the bearers of truth in inquiry, and it is beliefs that constitute the basis for habitual action in the world of truth (the 'real' world). "Inquiry," Forster states in paraphrasing Peirce, "is senseless unless there is a difference between truth and falsehood and, moreover, between thinking one is right and being right" (p. 162).

We can further unpack what Peirce means by the 'truth' by following Forster's (2011) four-part classification: correspondence theory of truth, coherence theory of truth, consensus theory of truth, and instrumentalist theory of truth. Before we do so, we emphasise, according to Forster (2011), that "these seemingly disparate conditions of truth are interdependent – that is, none can be realized unless each of them is" (p. 161).

It all starts for Peirce with the *correspondence theory of truth*:

> ...there is a way the world is, how it is does not depend on how inquirers think it is and that a belief is true if and only if it represents this reality." (Forster, 2011, p. 161)

Thus, for Peirce, something (a belief in symbolic form) is true only if it directly connects with that which is independent of us, what Peirce calls "some external permanency" (CP 5.384). There is a correspondence between reality and the truth. However, as Forster points out, Peirce's

---

[2] In referencing Peirce's work in this chapter, we use his *Writings* whenever his *Collected Papers* don't suffice. In doing so, we use the standard protocol: Wx: y = *Writings of Charles S. Peirce*, volume *x*, page *y*.

notion of correspondence is not the 'standard' notion. It is of a particular form that prioritises experience:

> ... the indexical connection of a symbol to its object is dynamical and unmediated ... experience does not separate knowers from their objects ... it is rather a point of contact at which symbols (cognition) and their objects (the world) meet. (Forster, 2011, p. 163)

Something can only be true if experience affords correspondence with reality. But what of the nature of the signs themselves; surely, they have a role to play?

Indeed, they do, and so Peirce also endorsed the *coherence theory of truth*:

> Inference is the derivation of one symbol (a conclusion) from other symbols (its premises) according to logical rules, or 'leading principles', and since symbols determined to be true on the basis of inferences represent real objects, it follows, on Peirce's view, that inferential relations among symbols 'demonstrat[e] the existence of a reality'. (Forster, 2011, p. 164)

It is not enough for Peirce for the objects of reality to brutally impinge on signs (in particular symbols) to compel them to the truth; the components that constitute these signs must also be logically connected with each other (i.e., cohere). Therefore, the truth of a sign (in relation to reality) is both a priori and a posteriori in nature.

Does all this mean that truth in its relationship with reality is an affair of the individual? Not at all for Peirce, the scientist, who knew well the necessarily communal nature of inquiry. As Forster (2011) states about Peirce's *consensus theory of truth*, "... truth is defined in terms of what inquirers agree on ... truth is intersubjective – that what is true for one inquirer is true for all of them" (p. 167).

However, Peirce made very clear that the truth is in no way determined by a particular community at a particular point in time, rather the truth is an inevitable "one True conclusion" (CP 5.384). Forster helps us to understand the nuances of Peirce's position on communal truth:

> The ultimate opinion of which Peirce speaks is one which would be vindicated over the continuum of possible experimental trials it implies. Since the content of a belief is not limited to any collection of actual events, it exceeds the (finite) experience of any actual community of inquirers. (Forster, 2011, p. 168)

In this way, whenever a community of researchers (e.g., us) undertakes inquiry, it will never ultimately determine the truth–Peirce is concerned here not so much with particular actualities as with all possibilities. However, this epistemic state only makes it all the more important for these finite researchers in finite time to undertake inquiry according to the method of science that puts us on the path to the truth.

Thus far, Peirce's account of truth seems to operate in an idealistic vacuum that is devoid of the practicalities of science and inquiry more generally. But there is more, for as Forster (2011) points out, the *instrumentalist theory of truth* is the way by which Peirce asserts the importance of actions (of all kinds).

> ...inquiry involves the deliberate formation of beliefs and is directed at the attainment of a definite end, namely truth. Since the content of a belief (according to the pragmatic maxim) is given by certain habits of conduct, inquiry 'essentially involves self-control' and practical deliberation." (Forster, 2011, p. 170).

For Peirce, inquiry undertaken as the method of science is a process of deliberately establishing and refining beliefs, as habits of action, which align with the attainment of the ultimate end, the truth. In other words, "to characterize truth in terms of instrumental reliability is to examine the role of symbols as reliable guides to action" (Forster, 2011, p. 173).

The truth is thus a matter of *what we do* (instrumental) just as much as it is a matter of linking with *what is out there* (correspondence), *what connects our beliefs as signs* (coherence) and *what we agree to as a community* (consensus). In the great tradition of Peirce as the advocate for the epistemic power of diagrams (Stjernfelt, 2007), we represent this four-part account of truth in a diagram (Figure 6.1) that also celebrates Peirce's penchant for sketches. In this diagram, Reality, composed of Objects, is linked to Signs by Correspondence. The Signs are linked by Coherence and this network of Signs converges to Consensus. Investigation (as the method of science that Peirce advocates, the instrumentalist component of truth) is the deliberate and controlled linking of Objects with Signs and Signs with each other in a habitual (or law-like) manner. Investigation is the action that brings about Correspondence and Coherence that enables Inquiry to converge on Consensus. In presenting this diagram as an icon full of epistemic potential, we aim to reveal the structural relationships between Reality (as composed of objects) and Truth (as composed of signs), as well as the structural relationships between the correspondence, coherence, consensus, and instrumentalist aspects of truth.

However, even if we make use of such diagrams, the full meaning of truth as correspondence, coherence, consensus, and instrumentality is not fully apparent. In particular, if we want to explore the significance of this truth for science education and our practices as researchers and teachers who are committed to inquiry that makes a theoretical and practical difference. So, we now turn our attention from Peirce to Franklin, and more specifically to Franklin as depicted in Burns' film 'Benjamin Franklin'. In doing so, we follow the lead of Campbell (1995) who astutely pointed out the affinity between pragmatism, in particular Peirce, and Franklin:

> ...we can make connections with at least two major philosophical issues prominent in both Franklin's thought and in later pragmatism debates: the concern with using intelligence to advance human well-being, and the possible range of human experience. (Campbell, 1995, p. 745)

In doing so, we adhere to Peirce's semiotics to emphasise that Burns' version of Franklin is distinct from other representations of Franklin, and that all representations of Franklin are distinct from the 'real' Franklin. In other words, the sign is not the object. What we know as 'Franklin' is a collection and conglomeration of different signs. We also embrace Peirce's pragmatic maxim to frame the meaning of our watching of Burns' documentary.

> Consider what effects, that might conceivably have practical bearings, we conceive the object of our conception to have. Then, our conception of these effects is the whole of our conception of the object. (CP 5.422)

In this way, 'Benjamin Franklin' as the object of our conception has certain conceivable practical bearings for us as viewers/listeners, researchers and teachers, and our conception of these effects is constitutive of the whole documentary that is 'Benjamin Franklin'. We stress once again that it is the particular filmic object 'Benjamin Franklin' as created by Ken Burns that afforded us, in sometimes-paradoxical ways, fresh insights into the importance of Peirce's notion of truth for our own methodological practices. But, as we will show, this filmic version of Franklin will never be enough on its own to provide a full account of Peirce's ideas, for we need the full range of sign forms to make this possible.

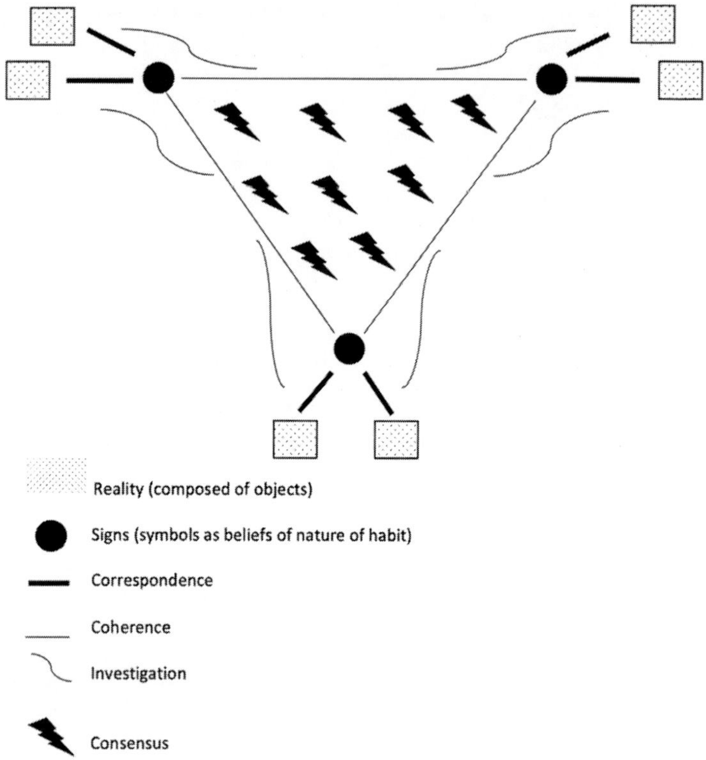

Figure 6.1: Peirce's four-part account of truth

Benjamin Franklin as the bearer of the truth

As with most of his documentaries, Burns visually presents paintings and prints along with verbalisations of written letters, proclamations, newspaper articles etc. to make 'Benjamin Franklin'. The use of narration and time-period appropriate music adds depth to this filmic portrayal of Franklin, as the viewer is immersed in an audio-visual landscape that conjures 18th Century England, France, and the U.S.A. during a time of revolution that would politically and scientifically define the Western world for centuries to come.

We propose that while Burns' version of Franklin positively personifies certain dimensions of Peirce's account of truth, in particular correspondence, coherence and instrumentalist aspects, Burns' portrayal of Franklin as a scientific genius who tamed and gifted electricity to humanity is at odds with truth as consensus. Therefore, while Immanuel Kant (1755/2012) did indeed describe Franklin as "the Prometheus of modern times", the way in which Burns presents this descriptor as defining Franklin (Figure 6.2) says more about the way in which Burns mythologises American individualism in his films than it does about Franklin the scientist.

With this in mind, we now provide examples of Franklin's scientific endeavours, as made apparent to us in Burns' documentary, and highlight the way in which these beliefs as habitual actions are instances of Franklin personifying the four different aspects of Peirce's account of truth.

Figure 6.2: The painting *Benjamin Franklin Drawing Electricity from the Sky*, by Benjamin West (ca. 1816), is presented onscreen by Burns and subjected to the 'Ken Burns effect' as the viewer is firstly zoomed in to Franklin's face and then panned to the key in his hand as Peter Coyote the narrator proceeds to paraphrase Kant in describing Franklin as "the modern Prometheus".

Correspondence

Burns clearly depicts Franklin as a scientist/statesman who was always trying to investigate what was 'out there'. Franklin is depicted on screen as driven by a need–almost an obsession–to understand that which is independent from us. Burns shows us that each time Franklin crossed the Atlantic Ocean as he travelled between the U.S.A. and Europe, he undertook systematic collection of data on the currents that enabled him to identify and map the Gulf Stream. Therefore, even when Franklin was in the midst of diplomatic activities that would make possible the American Republic, his attention was still fixed on determining a correspondence between the world of everyday existence and that which lies beyond. As Burns shows through his weaving of various visual and auditory material, this was always an endeavour driven by Franklin's experience of nature and the different relationships he established with this reality through his use of scientific apparatus to determine the truth.

Coherence

Burn's filmic version of Franklin is always carefully organising and orchestrating his thoughts and actions to maximise their epistemic impact, as evidenced in his various writings and speeches as a scientist and diplomat that Burns flashes on screen and directs to our ears. Burns takes particular care to emphasise that Franklin began his working life as a printer with the utmost respect for the power of the written word to change the beliefs of the American people as the basis for habitual action. In many ways, Franklin's working life was his whole life, a point that comes across in the way Burns structures his documentary as filmic argument to foreground this aspect of Benjamin Franklin the sign. Burns uses the constant voice of Mandy Patinkin as Franklin to convince the audience that the spoken words of Franklin cohered just as well as his written words. As the viewer, we are immersed in a cinematic world of Franklin writing and speaking with intelligence and humour to hold court with almost any crowd (royalty, diplomats, working men and women). This is Franklin as accomplished orator and writer who marshalled all the signs (icons, indices, symbols) at his disposal to form a logical web of meaning that formed the path to truth through/as inquiry. The truth coheres for Burns' Franklin.

Consensus

As we alluded to earlier, Franklin as depicted by Burns is not someone who invested in the collective of the scientific and diplomatic community, but rather he ventured out into the wilds of nature and the metropolis to determine the truth for himself. We suggest, once again, that this is most evident in Burn's decision to highlight Kant's famous prehumous epitaph for Franklin: "the Prometheus of modern times". Burns deploys an arsenal of visual and auditory devices to give cinematic form to Kant's words; Franklin 'discovers' electricity in various forms and in various ways (almost) by himself and bestows these insights onto the masses in the form of the lightning rod. The truth at play here is the result not of consensus but of individual genius.

However, we know that representations of Franklin other than those that constitute Burns' film depict Franklin as first and foremost a man of the community (scientific and political) for whom truth was only attainable by working with and for others. This is evident in various historical accounts of Franklin's involvement in the genesis of the Leyden jar. This important scientific instrument is the first capacitor–a mechanism for storing energy as charge at high voltage–which was initially developed by von Kleist and van Musschenbroek (of Leiden/Leyden) and developed by a number of scientists, before Franklin added lead sheets (and the term battery) before other scientists such as Aepinus developed the air condenser. The development of the Leyden jar and of our technology and concepts of electricity more generally were collective endeavours that built steadily rather than a descent from the mountain by the prophet. Franklin is also notable for not patenting his inventions and making his technological innovations freely available, as well as creating a fund to support young innovators. In this way, Franklin as committed to the community of inquiry is conspicuously absent from Burns' depiction of Franklin as the American Prometheus, a depiction important for the ideology of American individualism that emerged in the 19th century. And, in fairness to Burns, he does powerfully portray the way in which Franklin disowned his son (William Franklin) to ensure the success of the new republic as a geopolitical entity for the people (i.e., the community). As part of this filmic depiction, Burns makes sure to direct the viewer's attention to the significance of Franklin's famous remark, "A republic, if you can keep it."

However, it is worth noting, that even those representations of Franklin that do acknowledge his commitment to truth as consensus, do so by framing such communities as particular groups that operate in finite time, which is a contrast to the infinite community of possibilities that is the pragmatist notion of consensus as the basis for the truth. By neglecting consensus, this film, in contrast to other representations of Franklin which foreground his communal spirit, highlights by its absence the importance of consensus for Franklins' successes. Thus, it is not just the correspondences of Burns' representation of Franklin to Peirce's ideas but also the divergences that enable us to appreciate Peirce's ideas more fully.

Instrumentalist

In Burns' documentary, Franklin dedicates his life to the method of science to enact inquiry in order to understand as much as he can about the natural world in which he is immersed. Burns bombards the viewer with images and sounds that convince them that whether in the laboratory, home study or meeting halls of Europe, Franklin desires to always refine his beliefs as based on evidence to change his actions in habitual ways that enable increased understanding of the objects of reality. Burns invests significant time and cinematic energy in showing the way that Franklin created a list of actions that he aspired to achieve, as he undertook a disciplined and controlled process of realising/not realising these actions on this list each day to hold himself accountable. This version of Franklin as instrumentalist in his relationship with the truth–a picture of self-awareness and self-discipline–is also reflected in the way that Burns shows Franklin the scientist using scientific instruments that he created for undertaking inquiry. This method of science led to other inventions that changed the daily practices of the public in useful ways (e.g., bifocals, circulating stove, lightning rod).

Pursuing the truth in science education research

We propose, inspired by Bakhurst (2023), that our task is to build bridges between education and philosophy. There is an urgent need for reciprocity. Equally philosophers need to be more pragmatic and concerned with what their ideas can do for people, in the world. Our project is an exemplar of this reciprocity: teachers face challenges teaching science; Peirce furnishes us with philosophical ideas about the nature of truth as the ultimate aim of scientific inquiry; by applying these ideas we can strive to improve our teaching and generate richer evidence to critically evaluate the philosophical underpinnings of our methodological practices.

So, it is now time to ask ourselves: what are the methodological implications, for our research and teaching of science education, of embracing Peirce's account of truth? In the form of Franklin on film (in particular Burns' documentary, together with Franklin as depicted in countless other representations), we have a personification of this truth that we can look to and listen to in order to answer this question. We need to frame our science education research as the method of science in action as/through signs in service of the truth, but what does this involve in practice?

First, as we engage with the learning and teaching of science in classrooms and other sites of learning, we need to always remember that such processes are grounded in phenomena that are external/independent from what it is that we are experiencing. In other words, the inquiry we undertake as research is determined by 'what is out there'. The very process of scientific inquiry that teachers and students undertake as they enact science, ought to be treated as processes that make possible the understanding of the natural world as an external realm. This is truth as correspondence. In our own research, this means that we need to consider teaching and learning as knowable objects of inquiry (see, for example, Tytler et al., 2020).

Second, neither the inquiry we undertake as researchers nor the inquiry that students and teachers enact in the science classroom is aligned with the method of science, unless the various representational forms that constitute these processes are arranged in a coherent way. In this way, our beliefs and those of students and teachers in doing science, must be considered as ways of acting that are not only habitual and under our control, but also forming a logical constellation of meaning making as inquiry. It is imperative that all involved in science education appreciate that to enact the method of science is to orchestrate signs (icons, indices, symbols) in such a way as to crystalise a logic of meaning that is able to correspond to the objects of reality that we aim to connect with through minds-on and hands-on science. This is truth as coherence and instrumentalism. In our own research, this means that the objects of inquiry are only knowable if we make considered and systematic use of the full range of sign forms to act in ways that generate the observations, hypotheses, and conclusions that determine the inference-making and inference-testing that 'reveal' reality as the truth (see, for example, Ferguson et al., 2021).

Finally, while we necessarily must collaborate with each other and with both teachers and students as distinct communities of researchers who share and critique ideas to advance inquiry, we must not do so with the intent of determining anything as absolute. We must simultaneously value what we do as finite collectives as we acknowledge that this is in service of something much greater than us. In our own research, this means that we must strive with our colleagues and collaborators to understand as much as possible about the knowable objects of inquiry, but do so with humility (see, for example, Hannigan et al., 2016) in that we must recognise that we (as discrete individuals and communities of individuals) will never know everything, and even if we did we would not be aware of this final epistemic state. Therefore, in terms of our research-informed teaching in science education, we ought to position the truth as central to what we do and so focus our instruction on instilling the method of science as the true path of inquiry in our pre-service teachers, in-service teacher colleagues, and research students. This is truth as consensus.

We finish with a bold proposal: both our research and teaching ought to be inquiry as/through signs to realise the truth as the ultimate aim, however without ever getting there. As researchers and teachers, we could do worse than emulate Franklin's commitment to the pragmatist truth. In undertaking our science education research in order to understand and improve practices of teaching and learning, we ought to align with Franklin's way of doing science as a method that is committed to the truth and to enabling us to refine our habits of changing the world in fulfilling ways. And it is Burns' version of Franklin, which brings to life this Prometheus of the 18th century, that makes apparent the potentially productive consequences of doing so for our methodological practices.

Film, with a century and a half of experimentation, has much to teach us. The conjunction and disjunction of image, text, and sound that filmmakers have pioneered and refined can productively be employed in educational research. We propose that Burns' film is an exemplar of the way that film as a semiotically rich system of audio-visual signs can, together with more traditional representational forms of research, expand our methodological horizons through Peirce's semiosis. In coming to understand and appreciate Peirce's ideas, in particular about truth, we need to engage with not only Burns' representationally rich and dynamic version of Franklin but with an array of other representations of Franklin that make apparent different aspects of Franklin the scientist. While Burns' film is an imperfect representation of Peirce's notion of truth, this imperfection allows us to see and hear beyond the content of Burns' film to focus on the representational modes at play, and thus to expand our own representational modes of meaning making to invigorate our own methodologies. As such, we encourage you to read Peirce, watch 'Benjamin Franklin' and to immerse yourself in the world of cinema not only as an entertainment but as methodology.

References

Bakhurst, D. (2023). Philosophy's blindspot. *Aeon.* https://aeon.co/essays/education-should-matter-to-philosophy-what-took-so-long

Campbell, J. (1995). The pragmatism of Benjamin Franklin. *Transactions of the Charles S Peirce Society, 31*(4), 745-792.

Cripps Clark, J., & Ferguson, J. P. (2020). A return to cave painting? Doing education research as film-philosophy. In P. White, R. Tytler, J. Ferguson, & J. Cripps Clark (Eds.), *Methodological approaches to STEM education research: Volume 1* (pp. 50-69). Newcastle upon Tyne, UK: Cambridge Scholars Publishing.

Ferguson, J., & Cripps Clark, J. (2021). Where the truth lies: Peirce through the lens of the third man. In N. Barnes & A. Bedford (Eds.), *Unlocking social theory with popular culture: Remixing theoretical influencers* (pp. 33-46). Cham, Switzerland: Springer.

Ferguson, J. P., Tytler, R., & White, P. (2021). The role of aesthetics in the teaching and learning of data modelling. *International Journal of Science Education, 44*(5), 753-774.

Forster, P. (2011). *Peirce and the threat of nominalism.* Cambridge University Press.

Hannigan, S., Raphael, J., White, P., Bragg, L. A., & Cripps Clark, J. (2016). Collaborative reflective experience and practice in education through self-study and arts-based research. *Creative Approaches to Research, 9*(1), 84-110.

Kant, I. (1755/2012). Succinct exposition of some meditations on fire. In E. Watkins (Ed.), *Kant: Natural science* (pp. 309-326). Cambridge University Press.

Lane, R. (2018). *Peirce on realism and idealism.* Cambridge University Press.

Legg, C., & Hookway, C. (2021). Pragmatism. *The Stanford Encyclopaedia of Philosophy.* https://plato.stanford.edu/archives/sum2021/entries/pragmatism/

Peirce, C. S. (1931-1935). *Collected Papers of Charles Sanders Peirce (Volume 5)*, C. Hartsthorne & P. Weiss (Eds.). Harvard University Press.

Peirce, C. S. (1984) *Writings of Charles S. Peirce – A Chronological Edition*, E. C. Moore, M. H. Fisch, C. J. W. Kloesel, D. D. Roberts & L. A. Ziegler (Eds.). (Volume 2). Indiana University Press.

Stjernfelt, F. (2007). Diagrammatology: An investigation on the borderlines of phenomenology, ontology, and semiotics. Springer.

Thelan, D. (1994). The movie maker as historian: Conversations with Ken Burns. *The Journal of American History, 81*(3), 1031-1050.

Tytler, R., Prain, V., Ferguson, J. P., & Cripps Clark, J. (2020). Methodological challenges in researching, teaching and learning student capabilities: The "creativity" case. In P. White, R. Tytler, J. P. Ferguson, & J. Cripps Clark (Eds.), *Methodological approaches to STEM education research – Volume 1* (pp. 1-20). Cambridge Scholars Publishing.

Joseph Paul Ferguson (BSc, BA(Hons), MTeach, PhD) - Deakin University, School of Education, Faculty of Arts & Education, 221 Burwood Highway, Burwood, 3125, Victoria, Australia.

Joseph Paul Ferguson is an educational researcher and teacher educator interested in investigating the various manifestations of reasoning in the science classroom, in particular creative reasoning. Through an exploration of the philosophies and theories that underpin inquiry practices and the associated methodologies that enable this research, Joseph seeks to add to efforts to clarify what it means for teachers and students to enact inquiry in the classroom. Joseph is particularly interested in the affordances of video-based methodologies, informed by film theory and philosophy, to maximise the potential of such research to support teachers and students in their teaching and learning of science.
https://orcid.org/0000-0002-0971-3256

John Cripps Clark (BSc, DipEd, MAppSc, PhD) - Deakin University, School of Education, Faculty of Arts & Education, 221 Burwood Highway, Burwood, 3125, Victoria, Australia.

John Cripps Clark is a science and science education teacher and theorises and researches STEM education, online learning, cultural-historical theory, visual-linguistic reasoning, the pedagogy of judging the accuracy and reliability of information, school gardens, overseas professional experience, science games, Pragmatism, and film. For the last twenty years, John has run a cultural-historical & activity research reading group with members across the Asia-Pacific region. He is a coordinator for the Australian Association for Research in Education special interest group: Cultural-Historical & Activity Theory, and has organised four Asia-Pacific ISCAR-AARE Summer Schools.
https://orcid.org/0000-0002-0335-6964

CHAPTER 7

BECOMING ENTREPRENEURIAL ON OUR TERMS: CONCEPTUALISING ENTERPRISING RESEARCH METHODOLOGIES

CARLY SAWATZKI & JILL P. BROWN

Abstract

Our work as teacher educators and educational researchers takes place at a time of neoliberal university reform and in a field characterised by capitalist thinking (financial capability and financial education research). Beyond the traditional activities of teaching, research, and service, academics are under increasing pressure to operate as entrepreneurial public intellectuals who disseminate knowledge widely. In this context, this chapter explores our experience leading the *Economics + Maths = Financial Capability* project–a first-of-its-kind educational design research initiative designed to explore the affordances of an alternative vision for financial education at school, and to identify actions that key organisations can take to better support teachers as financial educators. The intervention we designed and tested was the *Economics + Maths = Financial Capability* course–a series of 6 x 90-minute seminars aimed at boosting teachers' curricular, content, and pedagogical content knowledge for teaching students about finance in Years 7-10. The course was offered on the free market via the School of Education's professional learning hub, PLEdHub@Deakin. Over the course of the project, we interacted with stakeholders across the education community via marketing, media, and public relations activities. In this chapter, we conceptualise these activities as enterprising research methodologies, because they ultimately helped us to achieve our research aims. We also problematise the pressure academics face to compete, produce, and perform in the modern academy. We conclude by reflecting on the legacy of the research experience on our academic values, identities, and work practices.

Key words: Financial capability, Financial education, Neoliberalism, Enterprising research methodologies, Academic identity

Introduction

For several decades, Western universities have been subject to wide-ranging structural, financial, and ideological changes (Barrow et al., 2022). Consecutive Australian governments have pursued neoliberal university reform, where resources are constrained, and scholarly activity is an increasingly competitive enterprise. Newly ascendant commercial and accountability imperatives place academics under the logic of new managerialism (Clegg, 2008), which privileges decision making based on market principles. In this climate, politicians and university leaders laud the potential of innovative applied research for industry alliances, income creation, and benefits to economy and society (Blackmore, 2020). At the same time, academics lament that pressure to become entrepreneurial is not only at odds with the scholarly tradition but has also compromised the ability to maintain independence and criticality (Antoniadou & Quinlan, 2021; Woollard et al., 2007).

While the remit of educational researchers is to generate and share robust empirical insights and design rigorous interventions, our ability to do so is increasingly at the discretion of power brokers and decision makers who define system and school priorities, as well as how these should be addressed. Declining research funding has left many academic careers in jeopardy, because developing new projects in their field is no longer feasible–marketisation has stymied free intellectual pursuit (Antoniadou & Quinlan, 2021) and is forcing the theoretical to be abandoned in favour of the commercialisable (Fuller et al., 2019). This means that academic autonomy has to be realised in new ways, by "managing multi-modality and multiple relationships" (Henkel, 2005, p. 173) with non-academic others, including funders and industry.

In this chapter, the example in focus is financial education research. We will examine our recent experience leading an educational design research project with funding from a community grants program. In particular, we will explore the tensions we confronted designing what was essentially a start-up initiative that we hoped would prove to be a viable and scalable teacher professional learning program aimed at improving the conditions for school students to develop financial capability.

Neoliberalism in the academy

The Oxford English Dictionary defines an academic as “a teacher or scholar in a university or other institute of higher education” and a scholar as “a specialist in a particular branch of study” (Oxford University Press, 2023). Academic work is understood to involve the pursuit of knowledge, as well as its dissemination and application. Like other universities, Deakin applies a workload allocation model to delineate academic work and measures performance against metrics. The workload allocation model categorises academic work as teaching, research, and service, and defines typical job tasks together with ‘reasonable estimates’ of how much time should be assigned to each task. Short and long-term career goals must be written, and the extent of their achievement assessed as part of annual performance reviews. Research expectations are set, and publications and grant income generated are recorded and allocated points. Research outputs matter, as they determine what percentage of your workload is allocated to research activity (produce more research outputs, be rewarded with more time for research activity). To be considered for promotion, academics are required to demonstrate sustained achievement of Minimum Standards for Academic Level. This includes evidencing that work activities align with Deakin’s strategic priorities and the needs of its communities. Archer (2008) and Sutton (2015) describe these managerial practices as newly penetrating forms of individualism and performativity. Davies (2005) describes the situation as relentless surveillance, where many academics report feeling value conflict or incongruence, as well as negative emotions including stress and exhaustion (Franco Santos et al., 2017), fear (Chubb et al., 2017), anger and frustration (Antoniadou & Quinlan, 2021), and feeling out-of-place (Gill, 2010). People who are recognised as successful in the academy tend to maximise their time and energy spent on activities that ‘count’–like writing grant proposals and articles for prestigious international journals (Mathies et al., 2020).

Beyond the traditional activities of teaching, research, and service, academics are under increasing pressure to be more entrepreneurial and public in their intellectualism than previous generations of scholars. Universities benefit reputationally by employing academics who are visible, known, and trusted to offer informed commentary on important issues, via a range of outreach and communication activities, including marketing, media, and public relations. While these sorts of tasks do not accrue research points, there seems to be the expectation that they will be undertaken within the time allocated to research activity. It's a performative form of job expansion and intensification for academics, and one that requires a different skill set to traditional academic work. In this chapter, we identify outreach and communication activities that proved essential to the success of the *Economics + Maths = Financial Capability* project and conceptualise these as enterprising research methodologies. We describe and discuss the process of becoming entrepreneurial through this set of systems and methods, noting that it was these activities that made it possible to generate the support of key stakeholders in completing the project. We examine the affordances and tensions associated with trying to gain support for critical and imaginative ideas in the public domain. We draw on literature exploring academic values, identities, and work practices, and situate our discussion in the context of our recent research experience.

Defining academic identity

Academic identity refers to the experiences, ideas, beliefs, and values of academics regarding the core values and principles that guide academic work (Tülübaş & Göktürk, 2022). In his seminal work on identity, Madan Sarup (1996) suggests two broad, constructionist, groupings for identity work. The first is the traditional view, which holds that sociological mechanisms such as class, gender, and race "operate simultaneously to produce a coherent, unified, fixed identity" (p. 14). The second, more recent view, emphasises the way that identity is "fabricated, constructed, in process" (p. 14) and is a composite of psychological and sociological elements. Here, identity is "necessarily incomplete, unfinished" (p. 47). Sarup points out that these views can overlap, since there can be "psychological aspects to the sociological factors, and sociological aspects to the psychological factors" (p. 14). The changing economic and political environment within which universities operate is transforming the structures and cultures within which academics work. In this context, Davies (2005) describes neoliberalism as reshaping academic identity, and academics as [potentially] "neoliberal subjects" (p. 1). She finds neoliberalism to be the source of jeopardy for academic identity, particularly for the life of critical and imaginative intellectual pursuits that are not perceived to respond to market demands. In this context, Tülübaş and Göktürk (2022) describe the enterprising academic as one who will do whatever it takes to be identifiable, measurable, comparable, and viable as declared by neoliberal norms.

Contextual background: Financial capability and financial education

Internationally, financial capability discourse tends to be characterised by capitalist thinking. In Australia, this is in part due to the Federal government's desire for its significant investment in this area to be tightly managed by government agencies with responsibility for the country's financial affairs. Until recently, consecutive National Financial Capability Strategies (Australian Government, 2022) were overseen by the financial regulator, the Australian Securities and Investments Commission (ASIC). Such oversight has now shifted to the Australian Government Treasury, the ministerial department responsible for providing the government with economic analysis and policy advice. It's a somewhat unique arrangement for these organisations to be controlling the purse strings and policies associated with an aspect of school education, and one that has been subject to academic critique (see Blue et al., 2014).

In the aftermath of the 2008 global financial crisis, ASIC lobbied for finance-related content to be prioritised as part of the new Australian Curriculum. The previous version of the Australian Curriculum (v8.4) frames opportunities for students to learn about finance within the Humanities and Social Sciences (where Economics & Business is taught and learned, and 'Consumer and financial literacy' is one of the key ideas) and Mathematics (where the 'Number and algebra' strand includes 'Money and financial mathematics' as a sub-strand). ASIC also created MoneySmart, a website to help people "take control of their money with free tools, tips and guidance" (see moneysmart.gov.au). Based on the assumption that explicit curriculum framing and a set of standardised lessons would ensure all students had access to financial education, MoneySmart was promoted as the mainstay teaching package for schools.

The new Australian Curriculum (v9.0) brings revised requirements. Economics & Business is offered in Years 5-10 and retains a focus on 'Consumer and financial literacy' in Years 7-10. However, the subject may be offered as an elective in Years 9-10 (Australian Curriculum, Assessment and Reporting Authority (ACARA), 2022b). In Mathematics, 'Money and financial mathematics' is no longer a sub-strand, but the expectation that students will learn about finance is made explicit in other ways. For example, the Mathematics curriculum emphasises "the importance of students learning how to apply mathematics to authentic real-world contexts" (see ACARA, 2022a) and makes explicit reference to teaching and learning about money and financial contexts within Achievement Standards from Years 2-10. The curriculum also specifies that students will learn to use mathematical modelling to solve practical and applied problems, including in financial contexts, when studying number, algebra, and measurement. We note that while it is possible to explore economics and finance when teaching statistics and probability, such opportunities have not been explicitly signposted and so may not be obvious to teachers.

Despite a place in the Australian Curriculum and a set of standardised lessons, schools typically view and enact financial education as a niche curriculum. Rather than feeling engaged and empowered, many teachers see finance as a special or taboo topic to be explored cursorily for one hour, one day, or one week each year. The default 'solution' continues to be large-scale quick fixes that are not necessarily informed by educational research, delivered by qualified teachers, or rigorously evaluated. Today, a growing number of taxpayer funded and commercial financial education programs promise to 'ease the burden' on 'overloaded' and 'time poor' teachers, and level the financial playing field for students. Government agencies, businesses, and community organisations continue to offer hubs of downloadable lessons and outreach incursions and excursions. More recently, millennial start-ups, social enterprises, and apps involving gameplay have begun to emerge. The market is looking crowded, and while there's no shortage of options for school leaders and teachers to choose from, teachers still report a lack of knowledge and confidence, and frustration with resources that aren't keeping up with the financial times.

We argue that the issue is a lack of strategic investment in teacher professional learning to show teachers how to design and implement better programs and lessons that can meet students' learning needs and interests. To begin to address this gap, we received funding from a community grants program for an educational design research project. Over the course of the *Economics + Maths = Financial Capability* project, we interacted with stakeholders across the education community via national symposia, teacher conferences, and media activity. New insights were generated through surveys and interviews with education professionals and secondary school teachers and students (parents and caregivers were beyond the scope of the project). The intervention we designed and tested was the *Economics + Maths = Financial Capability* course–a series of 6 x 90-minute seminars aimed at boosting teachers' curricular, content, and pedagogical content knowledge for teaching students about finance in Years 7-10. The course was offered on the free market via the School of Education's professional learning hub, PLEdHub@Deakin. It explored economic and mathematical concepts within contemporary financial contexts, including 'tap n go' technology; music, entertainment, and gaming subscriptions; buy now pay later and payday loan services; and climate-conscious consumption. The course was essentially a start-up initiative that we hoped would have longevity as a program of professional learning and research. Attracting teachers to the course and participants to the research stretched us into entrepreneurial conversations and activities–including marketing, media, and public relations–that we came to identify as enterprising methodologies.

Defining enterprising research methodologies

We define enterprising research methodologies as the marketing, media, and public relations activities that inevitably underpinned the success of the project. Our ability to recruit teachers to the course and participants to the research hinged on doing these things well. However, by forcing us to become entrepreneurial, the project also challenged our academic values, identities, and work practices. In this section, we identify and explain the interrelated 'what's and how's' so as to unpack the specialised nature of the tasks and somewhat taken-for-granted expansion and intensification of traditional academic work.

Knowing your value and naming your price

Early in the project, a decision needed to be made about the value and price of the course. An important first step was to clearly articulate our value proposition–what were we offering that others were not? The course being designed by teacher educators and educational researchers recognised as experts in financial education, together with current, topical classroom teaching ideas and resources mapped to the Australian Curriculum would instil trust in the quality of the design and content. The course being aligned with the Australian Professional Standards for Teachers would assure teachers that this professional learning would be recognised as part of the 20 hour per annum requirement to maintain registration to teach. The opportunity to study online and complete independent, live, and small-group activities from the comfort of your home or school offered safety and convenience during the pandemic.

When it came to pricing the course, we sought advice from the project Steering Committee, which included colleagues from the funding organisation and teacher associations. These colleagues drew on their experience offering webinars, short courses, and conferences. They advised us that free professional learning tends to attract registrations, but engagement diminishes over time. They believed that a reasonable and competitive price would convey the value of the course and attract a committed cohort who would sustain their engagement. To decide on pricing, we did two things. First, we estimated the costs associated with our design, delivery and administration, and divided the total by a minimum viable intake of 20 people. On this basis, break-even would exceed $1000 per person. Next, we researched the price of other short courses. We were particularly interested in equivalent duration courses designed by teacher educators and educational researchers, backed by university branding. We found examples ranging from $750 to $1250.

Having articulated our value proposition and run the numbers, we believed that we were offering a premium course. However, we wanted the course to be affordable. Hence, we made the decision to price the course at $350 per person, with a 15% discount for Deakin alumni and partner teacher association members. While our grant enabled us to justify pricing the course below market rates, we understood that commercial realities would dictate our ongoing viability beyond the funding period.

Choosing your words and spreading them widely

Communication activities were necessary to generate awareness about the course and attract registrations. However, different forms of communication necessitated the mastery of different genres of writing. For example, we had to develop short and extended versions of promotional text that could be used for direct email marketing, course flyers, and social media posts. Since academics are trained in complex thinking and expression, developing a sales pitch felt uncomfortable. One challenge we confronted was how to gain the attention of a busy target audience. A related challenge was the need to meet limited word counts. When writing text for direct email marketing and course flyers, a punchy title, bold headings, concise dot points, and a clear call to action aims to cut through a busy recipient inbox and ensure your message is opened, read, and acted upon. Social media posts were another way to spread the word about the project. Writing Tweets and choosing handles and hashtags didn't come easily. At times, it seemed the less words available to us, the more time we spent drafting. Significant time was spent brainstorming clever wording that would play on financial terms before phrases like “Financial education in schools is short-changing young people” and “Investing in teacher knowledge pays dividends” were eventually shared.

Our wordsmithing also extended to pitching conference presentations, preparing media articles, and recording podcasts. We had the opportunity to promote the course via ‘taster’ sessions at teacher association conferences. As these were hosted online due to COVID-19, we had to craft session titles and abstracts that would entice teachers to opt in. The aim was to generate targeted leads who would register for the course and/or refer a colleague to do so. Writing articles and recording podcasts for mass and teacher media outlets helped us to communicate with a large audience. We came to see media engagement as part of our remit to educate about important issues in education. *The Age* is the most-read Victorian masthead, with the digital edition reaching more than 5 million readers (Bachelard, 2022). Our article, “Time's ripe for financial literacy students can bank on” (Saffin et al., 2021) was readily accepted with minimal editorial intervention. The Australian Council for Educational Research’s *Teacher Magazine* reaches more than 60,000 email subscribers and achieves more than 2 million page views annually. Here, the audience is professional educators, so we were more able to discuss education policy, curriculum, and research insights. We wrote four articles and recorded one podcast interview for *Teacher Magazine* exploring the sorts of things schools can do to strengthen programs and connect with students’ experiences with fintech and interest in cryptocurrency. We had a very positive experience with the editorial team. It is hard to know who chooses to read media articles and listen to podcasts about financial capability and financial education, or the extent to which our insights resonate with the audience. However, education professionals and teachers routinely tell us, “I read the piece you wrote for …” and, “I listened to your podcast interview…”

Creating a stylised online learning experience

Being one of the first courses to be offered via the PLEdHub@Deakin meant experimenting with WordPress, an open-source website builder. While we were able to draw on our experience with similar platforms, choosing a colour palette, writing for the web, and sourcing images ensured a professional and stylish look and feel, and an easy-to-navigate learning experience for teachers. Significant time was invested getting the course website ‘just right’ and troubleshooting technical glitches. Our course website design serves as an exemplar for the PLEdHub@Deakin when tender applications are being prepared.

Growing our professional network

There were both strategic and opportune ways that we were able to extend our professional network over the course of the project. Our research assistant used Google and LinkedIn searches to identify organisations and individuals working in financial education. Often, contact details were in the public domain and/or could be gained via a cold call. Contacts were added to an Excel spreadsheet and invited to events that we hosted. They were also sent project updates, including links to our media articles. At the close of the project, our contact list consisted of more than 150 colleagues, including bureaucrats in State and Territory education and curriculum authorities. Of course, the teachers who completed the course and participated in the research were valued members of our professional network too. Many suggested the course to colleagues and wrote generous testimonials.

Event facilitation

We hosted two free national symposia for education professionals working to influence quality financial education at school via roles within government agencies, universities, not-for-profit organisations, and teacher associations. The first event was titled, "Financial capability: Questions of what and how and who" and the second, "Influencing a more inclusive approach to financial education". These events aimed to promote the critical exchange of knowledge and ideas via thought-provoking presentations by scholars of international influence, including First Nations colleagues. COVID-19 necessitated a Zoom format, but we were abundantly aware that online meetings were becoming tedious. Given our desire to reach and impact a high-profile and influential target audience, we engaged a values-based communication consultant to help with event planning and facilitation. More than a 'Master of Ceremonies', this consultant supported us to clarify our goals and hone our approach. Importantly, we were encouraged to set aside our PowerPoint slides and adopt a Q&A approach facilitated with journalistic flair. Behind the scenes, planning documents, coaching, and practice sessions prepared us to appear accessible, conversational, and unscripted. We were pushed outside our comfort zone, but experienced the benefits. Feedback from participants included, "That was the best Zoom meeting I have ever attended!"

Reporting

The first author had led financial education research consultancies that culminated in industry reports before. For example, a report that she prepared for the Australian Taxation Office is published on their website and has been accessed more than 2000 times. That is a significant audience who may be interested to read your next contribution. Hence, we wanted to be strategic in who we wrote for and how we presented our findings and arguments. Of course, we had reporting obligations to the funding organisation, but who did we really want to read and apply our insights? Ultimately, we made the decision to write the report for financial education policy and decision makers. We knew we had an engaged professional network who wanted to learn from our research and we felt that reaching an influential audience was more important than reaching a wide audience. We identified that the structure and style of the report would contribute to an enjoyable reading experience and perhaps lead to the document being shared. We examined reports by colleagues in similar and different fields of research as examples that might guide us. Where possible, we quizzed the authors to learn the rationale for particular presentation decisions. This confirmed that we were going to need expert help to create the right impression. We engaged two consultants whose support proved invaluable. The first was an editorial consultant who offered abundantly useful reactions and advice on a substantive draft. It is one thing to have deep knowledge of a field of research and project, it is quite another to clearly communicate what you have learned and why it matters! Architectural advice included that most power brokers will only read the Executive Summary, and that the recommendations and actions should be tabulated. We even received advice on what headings and subheadings would distil findings and arguments to the extent that the Table of Contents itself would tell the project story. The second consultant we engaged was a designer to help represent our data as effectively as possible. To do her job well, the designer needed a focused brief, including punchy text and precise statistics. Figures 7.1 and 7.2 are examples of infographics that we could not have created alone.

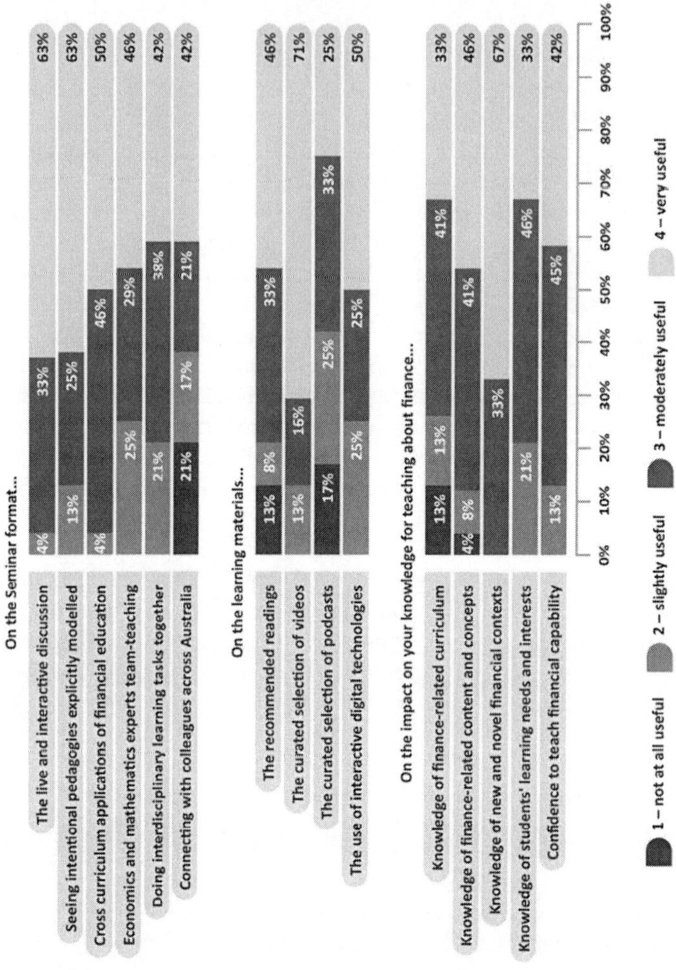

Figure 7.1: Teachers' Evaluations of the *Economics + Maths = Financial Capability* Course (n=24 secondary school teachers)

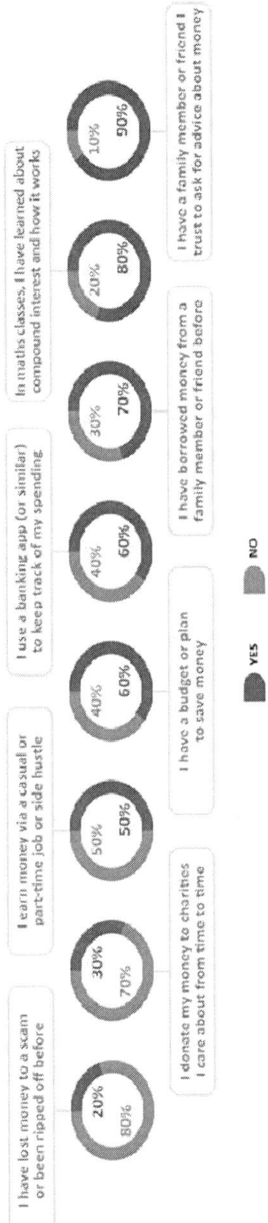

Figure 7.2: Students' Financial Activities (n=124 15-16-year-olds)

The challenge of cutting through to the masses

We were 'learning by doing' these enterprising research methodologies and our struggles were real. We had limited in-house professional support. Our project budget of < $100,000 was relatively small. We were working hard and wanted our work to be noticed and liked–and not just because this would be personally and professionally rewarding. We had to recruit teachers to the course and participants to the research if we were going to meet our contractual obligations to the funding organisation. Course registrations were a crude measure of the effectiveness of our marketing, media, and public relations activities. After all, our target audience was contending with the impact of the COVID-19 pandemic on their work and lives. At times, we felt invisible and disheartened. For example, we chose the quintessential pink piggy bank image to promote the course via social media, believing this would grab teachers' attention. We posted on 'hot' days and times when people are known to be scrolling their feed. Event reminders with modified messaging–"It's not too late to register!"–produced little effect. Analytics revealed our posts were being viewed–we simply weren't achieving the uptake we hoped for. This was stressful: devoting time to activities that did not generate the desired outcome (teacher registrations, research participants) came at a cost, and while the project design and timeline were able to be adjusted, the funding available was finite. While social media proved fickle, our professional networks were powerful referral sources. Forty-five percent of teachers who completed the course found out about it via a teaching colleague. One third registered via their teacher association. Around one in five had attended a course or conference with us before. These outcomes confirmed that our professional networks and relationships are invaluable, and real connections trump clickbait.

Tensions: Neoliberal 'push' vs scholarly 'pushback'

Education in Australia is such that schools are being positioned as belonging to a market, and it is assumed that market responses offer a reliable indication of what resources and programs are needed and valued (whether free or paid). Right now, marketisation means that the professional work of teachers–planning for learning and teaching lessons–is viewed as transactional and easily outsourced. Marketisation and standardisation go hand-in-hand. So, while educational research points to the need for local ownership and tailored approaches, the prevailing response is to 'standardise and deliver'. Narratives surrounding the teacher shortage and out-of-field teaching phenomenon are driving commercialisation across a range of curriculum areas. In the case of financial capability, taxpayer funded and commercial financial education programs claim to make it easy for teachers to deliver finance lessons. One stakeholder pointed out, "You're competing with The Barefoot Investor [NewsCorp journalist, Scott Pape] and his Money Movement Manifesto" (The Barefoot Investor, 2021). We are often approached to endorse Program X, Y, or Z as being suitable for all schools. The presence of well-resourced players vying for teaching and learning time is a stark reality. It is an ongoing issue that we find ourselves problematising and pushing back on these ways of framing teachers and teaching and diplomatically explaining that particular free and paid resources and programs may be saleable, but they are not educationally sound.

A related issue is that we are increasingly in competition with well-resourced corporate and market research consultancies who deal in flashy reports and slide stacks complete with large datasets, descriptive statistics, and images of smiling teachers and students. These organisations abide by different ethical standards in research. They can work fast, because they have in-house professional support, including callers, statisticians, and designers. We have written before about the tensions for us as teacher educators and educational researchers applying educational design research methodology as part of contract research initiatives (see Brown et al., 2022). Educational design research methodology draws on theory and empirical evidence to drive robust interventions which are studied via sophisticated data collection and analysis practices. Unlike corporate and market research consultancies, academics work in ways that are thoughtful, deep, and slow. Traditional academic work adheres to strict ethical standards in research, but writing and modifying ethics applications is time consuming. The work we do is hard, but important. These ways of being are central to scholarship, but at odds with the idea that research is a competitive enterprise, the value of which is best judged by performance indicators.

We set out to design a viable and scalable professional learning and research program, but we grew sceptical about the numbers and narratives that we were under pressure to deliver. Since it seemed to be assumed that achieving national reach and registering large numbers of schools and teachers to the course would in some way validate what we had to offer, it became necessary to find alternative ways to manage perceptions of the project. Research evidence and project data at the ready, we worked hard to construct a 'why' narrative that reflected a 'progress and learnings' orientation. As it is human nature to want to be regarded positively, such conversations were professional, but uncomfortable for us. Barrow et al.

(2022) describe inevitable identity questions for academics in how they 'go on' with their interactions with funders and industry. Implicit in the funding and reporting arrangement was an expectation to conform to neoliberal ways of working. Striving to compete, produce, and perform in ways that were in conflict with our values and the expectations associated with our workload allocation created identity tensions: Who am I, and what do I want my academic work to achieve?

Our experience was that it's very hard for scholarly design to compete on the free market. It's clear that the course offered something valuable–our research participants told us as much. In the absence of system and/or school level support, around half of the teachers who completed the *Economics + Maths = Financial Capability* course self-funded to attend and gave their personal time to engage with the professional learning opportunities. These are issues that we could not possibly tackle or address via our project design or enterprising research methodologies. Without system and school-level change that supports teachers to learn to be better financial educators, our sense is that large-scale quick fixes will continue to prevail.

Concluding remarks: Becoming entrepreneurial on our terms

Academic work will only continue to be transformed by neoliberal university reform. Financial capability discourse and financial education research will likely also continue to be characterised by capitalist thinking. This chapter has shown the expansion and intensification of our own work to include enterprising research methodologies requiring a different skill set to traditional academic activities. Billot (2010) describes that tensions arise for academic identity when there is deep incongruence between the values of individual academics and the field and organisations in which they work. We have identified the tensions associated with having to become entrepreneurial without having internalised the norms and discourses of neoliberalism. As these tensions are ongoing, so too academic identity formation is an unfinished project (Barrow et al., 2022). In financial education research, as in many other fields, it is a struggle to define who we are as academics and how we can best respond to what is happening around us. We are required to navigate neoliberal university reform, a rapidly changing financial world, and an emerging field of study. As teacher educators and educational researchers, we find ourselves caught between a neoliberal rock and a hard place, waiting for power brokers and decision makers to value and invest in what we offer. In the meantime, other educational priorities and projects demand our attention.

The *Economics + Maths = Financial Capability* project generated significant learning for all involved. For us, somewhat unanticipated was the extent to which the project stimulated new insights into our academic values, identities, and work practices. These insights emerged due to the tensions and discomfort we experienced feeling like entrepreneurs driving a start-up initiative. For example, we learned to engage in marketing, media, and public relations activities that could help us to achieve our research aims. Engaging an event facilitator, editorial consultant, and designer taught us to distil complex, sometimes opaque thinking into useable insights. Working with non-academic partners made us better public intellectuals.

The expectation to measure and report on traditional research outputs suggests arranging work time and practices according to 'what counts'. Inevitably, spending time on becoming entrepreneurial came at the expense of more traditional academic tasks, like writing the next grant proposal or writing for prestigious international journals. Whilst we felt being pulled in different directions was uncomfortable, Clegg (2008), and Tülübaş and Göktürk (2022) argue that academics *can* find ways to exercise principled personal autonomy and agency, despite the managerial context of the modern academy. Archer (2007) argues that people exercise reflexivity through considering their social contexts in relation to their subjective self and vice versa, and eventually determine their courses of action under the circumstances. Upon reflection, navigating these tensions was part of becoming entrepreneurial on our terms.

As a result of this research experience, we have achieved greater clarity on 'what matters' to us and how to prioritise or 'go on' with our work in financial capability and financial education, particularly when interacting with non-academic others. Specifically, the project helped cement five informed beliefs that can guide our future activities:

- As the financial landscape is dynamically changing, so too is what it means to be financially capable.
- Teaching about finance is an important part of the Australian Curriculum and teachers' work.
- The finance-related content and contexts that financial education programs should include will continue to evolve.
- Teachers' professional knowledge and work are highly specialised and constantly developing.

- We want to recognise teachers as professionals and build, not erode their knowledge and confidence as financial educators.

The project also strengthened the value we place on intellectual curiosity and thoughtful, deep, and slow discipline scholarship. To an extent, the collaborative reflective practice exemplified by this chapter, is testament to how we see ourselves as educational researchers. Educational design research sustains our work as scholars. It's not easy to do or measure, but it can make a difference to teachers and students.

References

Antoniadou, M., & Quinlan, K. M. (2021). Holding true or caving in? Academics' values, emotions, and behaviors in response to higher education reforms. *Higher Education Policy, 35*(2), 522-541.

Archer, M. (2007). *Making our way through the world.* Cambridge University Press.

Archer, L. (2008). Younger academics' constructions of "authenticity", "success" and professional identity. *Studies in Higher Education, 33*(4), 385-403. https://doi.org/10.1080/03075070802211729

Australian Curriculum Assessment and Reporting Authority. (2022a). *Revisions to the Australian Curriculum: Mathematics.* https://v9.australiancurriculum.edu.au/resources/stories/revisions-to-the-australian-curriculum-mathematics

Australian Curriculum Assessment and Reporting Authority. (2022b). *Understanding this learning area: Humanities and Social Sciences.* https://v9.australiancurriculum.edu.au/teacher-resources/understand-this-learning-area/humanities-and-socialsciences

Australian Government. (2022). *National financial capability strategy.* https://www.financialcapability.gov.au/sites/www.financialcapability.gov.au/files/2022-02/Financial-Capability-Strategy-2022_0.pdf

Bachelard, M. (2022, November 21). The Age retains its lead as the biggest Victorian masthead in the country. *The Age.*

Barrow, M., Grant, B., & Xu, L. (2022). Academic identities research: Mapping the field's theoretical frameworks. *Higher Education Research & Development, 41*(2), 240-253.

Billot, J. (2010). The imagined and the real: Identifying the tensions for academic identity. *Higher Education Research & Development, 29*(6), 709-721. https://doi.org/10.1080/07294360.2010.487201

Blackmore, J. (2020). The carelessness of entrepreneurial universities in a world risk society: A feminist reflection on the impact of Covid-19 in

Australia. *Higher Education Research & Development, 39*(7), 1332-1336. https://doi.org/10.1080/07294360.2020.1825348

Blue, L., Grootenboer, P., & Brimble, M., (2014). Financial literacy education in the curriculum: Making the grade or missing the mark? *International Review of Economics Education*, *16*, 51-62.

Brown, J. P., Sawatzki, C., & Widjaja, W. (2022). Tensions between theory and practice: A mathematics education perspective. In P. White, R. Tytler, J. Ferguson, & J. Cripps Clark (Eds.), *Methodological approaches to STEM education research* (Vol. 3., pp. 286-307). Cambridge Scholars.

Chubb, J., Watermeyer, R., & Wakeling, P. (2017). Fear and loathing in the academy? The role of emotion in response to an impact agenda in the UK and Australia. *Higher Education Research & Development, 36*(3), 555-568. https://doi.org/10.1080/07294360.2017.1288709

Clegg, S. (2008). Academic identities under threat? *British Educational Research Journal, 34*(3), 329-345. https://doi.org/10.1080/01411920701532269

Davies, B. (2005). The (im)possibility of intellectual work in neoliberal regimes. *Discourse, 26*(1), 1-14. https://doi.org/10.1080/01596300500039310

Franco-Santos, M., Nalick, M., Rivera-Torres, P., & Gomez-Mejia, L. (2017) Governance and wellbeing in academia: Negative consequences of applying an agency theory logic in higher education. *British Journal of Management, 28*(4), 711-730.

Fuller, D., Benyon, M., & Pickernell, D. (2019). Indexing third stream activities in UK universities: Exploring the entrepreneurial/enterprising university. *Studies in Higher Education, 44*(1), 86-110. https://doi.org/10.1080/03075079.2017.1339029

Gill, R. (2010). Breaking the silence: The hidden injuries of the neoliberal university. In R. Ryan-Flood & R. Gill (Eds.), *Secrecy and silence in the research process: Feminist reflections* (pp. 228-244). Routledge.

Henkel, M. (2005). Academic identity and autonomy in a changing policy environment. *Higher Education, 49*(1-2), 155-176. https://doi.org/10.1007/s10734-004-2919-1

Mathies, C., Kivistö, J., & Birnbaum, M. (2020). Following the money? Performance-based funding and the changing publication patterns of Finnish academics. *High Education, 79*(1), 21-37. https://doi.org/10.1007/s10734-019-00394-4

Oxford University Press. (2023). Oxford Languages. www.oed.com

Saffin, P., Sawatzki, C., & Brown, J. (2021, November 7). Time's ripe for financial literacy students can bank on. *The Age*.

Sarup, M. (1996). Identity*, culture and the postmodern world.* University of Georgia Press.
Sutton, P. (2015). A paradoxical academic identity: Fate, utopia and critical hope. *Teaching in Higher Education, 20(1), 37-47.* https://doi.org/10.1080/13562517.2014.957265
The Barefoot Investor. (2021). The money movement manifesto. https://www.barefootinvestor.com/articles/the-money-movement-manifesto
Tülübaş, T., & Göktürk, Ş. (2022). A meta-synthesis on academic identity in the neoliberal context of academy. *Higher Education.* https://doi.org/10.1007/s10734-022-00906-9
Woollard, D., Zhang, M., & Jones, O. (2007). Academic enterprise and regional economic growth: Towards an enterprising university. *Industry and Higher Education, 21*(6), 380-477. https://doi.org/10.5367/000000007783099836

Carly Sawatzki (BA/BBus(Mgt), GradDipEd(Sec), MEd, PhD) - Deakin University, School of Education, 221 Burwood Highway, Burwood, 3125, Victoria, Australia.

Carly Sawatzki is a teacher educator and educational researcher in the School of Education. She has more than 15 years' experience working with preservice and practising teachers (including out-of-field and non-specialist teachers of mathematics) across primary and secondary courses. Carly's work is distinctly 'real world' and aims to gently influence the way teachers think about educating young people for active and informed citizenship. She is internationally recognised for her classroom research which explores how young people develop numeracy and financial capability within families, communities, and schools. Carly works collaboratively with teachers and students to design interdisciplinary learning tasks situated in contemporary financial contexts. Carly has published in prestigious international journals and led research and curriculum consultancies for Australian State, and Territory education authorities. She is regularly engaged by teacher associations and schools, being recognised as a thought-provoking presenter who draws on research to challenge thinking, promote critical conversation, and inspire innovation.
https://orcid.org/0000-0002-1463-8844

Jill P. Brown (BSc(Ed), GC(Ma&MaEd), MEd, PhD, GCHE) - Deakin University, School of Education, 221 Burwood Highway, Burwood, 3125, Victoria, Australia.

Jill P. Brown is a teacher educator and educational researcher in the School of Education. Jill has more than 20 years' experience in higher education, where she has focused on mathematics and mathematics education for preservice and practising teachers (including out-of-field teachers) in early childhood, primary, and secondary settings. She is the Course Director of Deakin's Graduate Certificate Secondary Mathematics for out-of-field teachers. Jill's research in the field of mathematical modelling, the teaching and learning of functions, and the use of digital technologies by teachers and students is internationally recognised. She is an elected member on the International Executive of the International Community of Teachers of Mathematical Modelling and Applications (ICTMA), a member of the expert panel for the International Mathematical Modelling Challenge (IMMC), and a member of the Australian IMMC advisory team and judging panel.
https://orcid.org/0000-0002-9202-3634

CHAPTER 8

ISSUE-CONCEPT MAPS AS A METHODOLOGICAL TOOL FOR EXPLORING STUDENTS' UNDERSTANDING OF SOCIOSCIENTIFIC ISSUES

GAHYOUNG KIM & KONGJU MUN

Abstract

The development of science and technology brings a prosperous life to humans, but it also causes social controversy due to the potential risks of such advancements. Socioscientific issues (SSIs) caused by science and technology are complex and controversial as they are connected at individual, social, and global levels and can involve ethical and moral issues depending on various stakeholders. Recently, as abnormal climate phenomena related to human-induced climate change, such as floods and bushfires, have become more severe and unpredictable in Australia, the importance of SSI approaches is increasingly recognised. Concept maps are widely used in education research to explore students' conceptual changes and alternative conceptions. Many researchers exploring SSIs still rely on using multiple-choice, true/false, or Likert-type instruments for investigating students' conceptual understanding of SSIs. In this chapter, we draw on our previous research of students' conceptual understanding of climate change and fine-dust issues through the process of a design-based research cycle to explore the affordances and limitations of Issue-Concept Maps (ICMs) for analysing students' conceptual understanding of SSIs. We link these methodological affordances and limitations to the teaching and learning strategies for using ICMs, particularly how they can be used when selecting SSI topics, constructing lesson contents, and assessing students' conceptual understanding of SSIs.

Key words: Issue-concept map, Socioscientific issues, Conceptual understanding, Climate change, Fine-dust

Introduction

Recently, climate change has become a significant social and political issue as large-scale, unexpected abnormal weather events such as floods, bushfires, and hail have occurred frequently in Australia. Unlike natural causes in the past, the leading causes of climate change have been human activities since the Industrial Revolution. Climate change causes the earth's temperature to rise due to human activities associated with affluent lifestyles, such as carbon dioxide emissions due to excessive energy use and natural damage due to indiscriminate development. This phenomenon is organically and intricately connected to the entire earth system, causing a crisis for humans and the earth's ecosystem. In global issues such as climate change, the subject of causality is only occasionally identified. As in the case of Tuvalu, which is losing its land due to sea level rise, the results of human activities can cause enormous damage to specific areas. In addition, since various stakeholders are related to the issue, it is necessary to approach global issues from social, economic, and political perspectives with an ethical-moral overlay.

Sadler and his colleagues (2007) defined Socioscientific issues (SSIs) as controversial social problems arising from science and technology and real-world problems that often have ethical components, such as climate change. In science education, the need for SSI instruction has emerged to enable students to make decisions and take actions as the agent of solving SSIs beyond understanding science-technology-society interrelations. SSI instruction can increase students' conceptual understanding of scientific content (Sadler & Zeidler, 2005) embedded in SSIs, appreciate the Nature of Science (Kolstø, 2001; Zeidler et al., 2002), and promote moral reasoning (Zeidler & Keefer, 2003), argumentation skills (Sadler & Donnelly, 2006), and reflective judgement (Zeidler et al., 2009). Also, SSI education can be a vehicle to promote students' moral sensitivity and empathy (Fowler et al., 2009), environmental awareness (Mueller & Zeidler, 2010), character (Lee et al., 2013; Kim & Lee, 2017), and sense of place (Kim et al., 2020) for considering SSIs with multi-perspectives.

One goal of SSI education is to improve students' understanding of SSIs. Some researchers have conducted tests with multiple choice or short answer questions (Boyes & Stanisstreet, 1997; Andersson & Wallin, 2000; Boon, 2010, 2014), Yes/No and multiple-choice tests (Boylan, 2008) to explore students' understanding of climate change and true/false knowledge-based tests with an interview (Karpudewan et al., 2015). However, these short-answer or multiple-choice surveys tend not to be very revealing in exploring students' understanding of climate change. To compensate for this, Dawson (2015) investigated secondary school student's knowledge of climate change with a written questionnaire, "What is the greenhouse effect?", "What is climate change?" and Boon (2016) included an open-ended question on the survey, "Climate Change is caused by..." Although these studies included open-ended questions that allowed students to show what they knew about climate change, even these methods are unlikely to elicit responses from students considering it from various perspectives. SSIs are organically and complexly intertwined with multidisciplinary elements of science, technology, engineering, and humanities and social sciences, as well as environmental, social, economic, and political factors (complexity), inherent potential risks of science and technology (uncertainty), and ethical and moral values because they can be interpreted from various perspectives when applied to society. Considering these characteristics of SSIs, understanding of SSIs should be accompanied by education on the cause, effect, and countermeasures at the personal, societal, and global levels of SSIs (Choi et al., 2012; Dove, 1996; Lee at al., 2012; Morgan & Moran, 1995).

Research background and context

We have developed and implemented various SSI programs in secondary schools. To understand students' understanding of SSI, multiple-choice, short-answer, and written tests were conducted like other researchers. However, these methods have limitations because they do not provide a detailed picture of students' understanding of controversial issues. Most students understand the causes and phenomena of climate change on a global scale, but they need to consider the impact of climate change on them and the impact of their actions on climate change more comprehensively. In other words, they tend to think that SSI is not relevant to them and only see SSI as a social or global problem. Therefore, we raised the questions of how students understand SSI issues in different contexts, how we can measure and what SSI instructions are needed to promote student's understanding of SSI issues.

We developed an Issue-Concept Map (ICM) based on these questions and applied it to SSI education several times. We plan to re-implement SSI programs with ICMs for various SSI topics and generations in the future.

To this end, we tried to refine the ICMs tool by reflecting on our previous studies and reconsidering the application of ICMs to SSI education through design-based research.

Design-based research focuses on particular learning problems and aims to produce a measurable change in student learning in the classroom (Anderson & Shattuck, 2012). Design-based research is implemented over a long period through repetitive educational experiments with designed instructional tools or interventions in the real-world classroom context. Compared to a one-time input, researchers are more likely to observe the overall effect of an instructional tool on student learning (Barab & Squire, 2004; Brown, 1992; Sandoval & Bell, 2004). Also, researchers can adjust the instructional tools they are using to account for the unexpected conditions in the learning environment (Collins et al., 2004). Design-based research involves multiple cycles of designing, testing, refining, and further testing in an authentic context (Anderson & Shattuck, 2012; Barab & Squire, 2004; Brown, 1992; Collins, 1992).

In this chapter, we introduce ICMs as one way to consider SSI characteristics. We analyse our previous studies on students' conceptual understanding of climate change and fine-dust issues through the process of a design-based research cycle. In particular, we reviewed the reflections on the designing and testing process of design-based research. We discuss drawing out the limitations as well as affordances of ICMs in SSI education and refining the ICMs tool.

ICMs, developed by extending traditional concept map structures, are classified into cause, effect, and countermeasure according to occurrences, and each occurrence is divided into personal, societal, and global levels according to context (Mun et al., 2014). In this context, the scope of the personal level refers to individuals or families, the societal level is a local community or country, and the global level refers to several countries or the globe. Like a conventional concept map, students write down the concept corresponding to each cell, mark the relationship between concepts with connecting lines, and write linking words to describe the relationship (Figure 8.1).

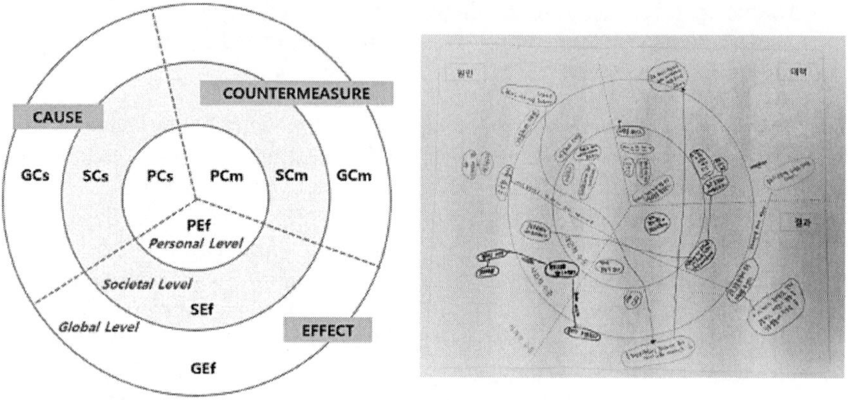

Figure 8.1: ICMs worksheet and student's artefact example (Kim et al., 2020)

Our first study (Mun et al., 2014) using ICMs investigated the understanding of climate change of 155 high school students (G10). Participating students completed ICMs in a group after completing 'Energy and Environment', a topic in the National Science Curriculum in Korea. As a result of this study, ICMs played a role in facilitating students to think about various ideas and concepts according to contexts and occurrences of climate change. In this study, ICMs were helpful in exploring students' understanding of SSIs and for students in expanding the connection between concepts into various patterns. We have confirmed that ICMs can be useful learning tools in SSI instruction.

Our second study (Kim et al., 2020) applied ICMs to measure the change in middle school students' understanding of the fine-dust issue, a local issue, rather than climate change, the global issue. Unlike the first study, we implemented the 'Fine-dust Zero' of community-based SSI programs for 16 lessons in 8 weeks. Students wrote ICMs for fine-dust before SSI class, and reported on the same ICMs sheet in different colours after class. As a result of the study, it was possible to explore how students' understanding of fine-dust issues changed using ICMs. In addition, ICMs helped identify students' prior knowledge of SSI issues and measure educational effects by planning and implementing SSI instruction.

However, there may be affordances and limitations on the use of ICMs depending on the subject and context of SSIs, and there may be an unclear boundary between information and concepts in written ICMs. We link these methodological affordances and limitations to the teaching and learning strategies for using ICMs, mainly how they can be used when selecting SSI topics, constructing lesson content, and assessing students' conceptual understanding of SSIs.

Methodological affordance 1: Selecting SSI topics

The first decision teachers face when planning the SSI lesson is topic selection. Zeidler and Sadler (2007) characterised SSI as complex, ill-structured authentic problems requiring moral reasoning and having indeterminate solutions. In the past 20 years, SSI education has dealt with global issues, such as climate change, genetically modified foods, nuclear power generation and waste, water shortages, infectious diseases, and marine plastics. However, these global issues may have different sensitivities depending on the learner's local context and situation. For example, in the case of climate change, Korean students who are not directly affected by climate change tend not to consider climate change as an issue that dramatically affects them, even though it is a problem at the global level. Another example is the issue of fine-dust in China that has emerged as a severe issue in Korea since 2015, affecting citizens' health and daily lives. However, fine-dust in other countries has never become a social issue in Australia.

In previous research, students' responses to the cause, effect, and countermeasures to the problem of fine-dust have provided specific answers at a personal level, which contrasts with their perception of climate change as a social and global issue. The fine-dust issue is more relevant and sensitive than climate change to Korean students in 2016. Controversial issues confined to a specific region can be considered at the personal and societal levels. For example, in Korea, the issue of abandoned animals has been an urban issue since 2015. Some researchers developed and implemented an abandoned animals SSI program for secondary students who live in Seoul (Kim, 2016) and explored the effect of students' understanding of SSI, character development, and sense of place after implementing this program (Kim, 2016; Kim & Lee, 2017; Kim et al., 2020; Kim & Lee, 2021).

Teachers must check the extent to which the local context and situation, such as specific issues of school region and students' thoughts, and so on, are relevant to the SSI topic. This is because SSI topics are multidimensional, and cause and effect may differ depending on the local context and situation. We, therefore, suggested that teachers should try to write their ICMs on the SSI topics before class, and if possible, consider a relatively narrow range of SSI topics that are locally relevant. Another approach is to first survey students' thoughts on broader issues, such as climate change. After doing this, you can move to local, specific issues, such as wildfires and flooding. You can also create new ICMs on topics that emerge from the ICMs that students create.

For example, in our 2014 study, high school students were asked to write ICMs on climate change. Students answered 'floods' and 'flood sufferers damage' as 'Influence' on climate change. Based on the results of this ICM activity, teachers can plan a follow-up lesson on the topic to allow students to discuss flooding in more detail. In particular, given the local context of the recent increase in severe flooding due to extreme weather events in South Korea, flooding may be an appropriate SSI topic choice for Korean students.

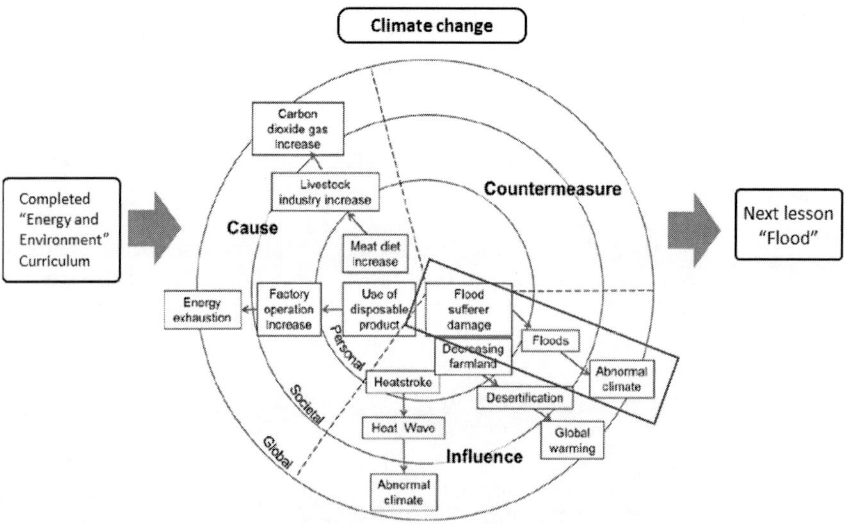

Figure 8.2: The strategy of selecting SSI topics (Drawing on Mun et al., 2014)

Methodological affordance 2: Constructing SSI lesson contents

Once an SSI topic has been selected, teachers must consider what content and activities to construct for the SSI lessons. The contents to be learned in SSIs are vast and complex, such as the convergence of various disciplines, complex relationships among multiple stakeholders, and moral and ethical approaches. In addition, learners' prior knowledge and experiences related to SSI may differ depending on their personal background or environment. Therefore, teachers need to assess students' understanding of topics in advance. Teachers can use ICMs as a diagnostic tool for constructing the content during lesson planning. ICMs guide students to think about issues in various contexts, not only on the global but also on the societal and personal levels, with ideas about the causes and effects of and countermeasures for issues. The teacher can visually grasp how the student responses of the pre-ICMs were distributed to nine cells, in which areas misconceptions appeared, and the relationship between concepts. In our study on fine-dust in 2020, the results of students' pre-ICMs on fine-dust are shown in Figure 8.3. First, teachers can diagnose that students believe the global cause of fine-dust is 'yellow dust from China', but they need help understanding why fine-dust is formed. Teachers can consider designing lessons that are helpful for students to understand the distinction between yellow dust occurring naturally and fine-dust caused by human activities. Second, the students' responses did not show the influence and countermeasures of fine-dust at the societal level. To address this, teachers can add activities to measure fine-dust in real-time in their community so that they can recognise the seriousness or seek to learn about what symptoms and diseases fine-dust causes when it enters the respiratory tract. Lastly, in the countermeasure against fine-dust, the students wrote general answers as a solution. Teachers can plan lessons to help students find viable solutions by adding activities for understanding relationships between various stakeholders involved in the issues and experimentation with face mask structure and function.

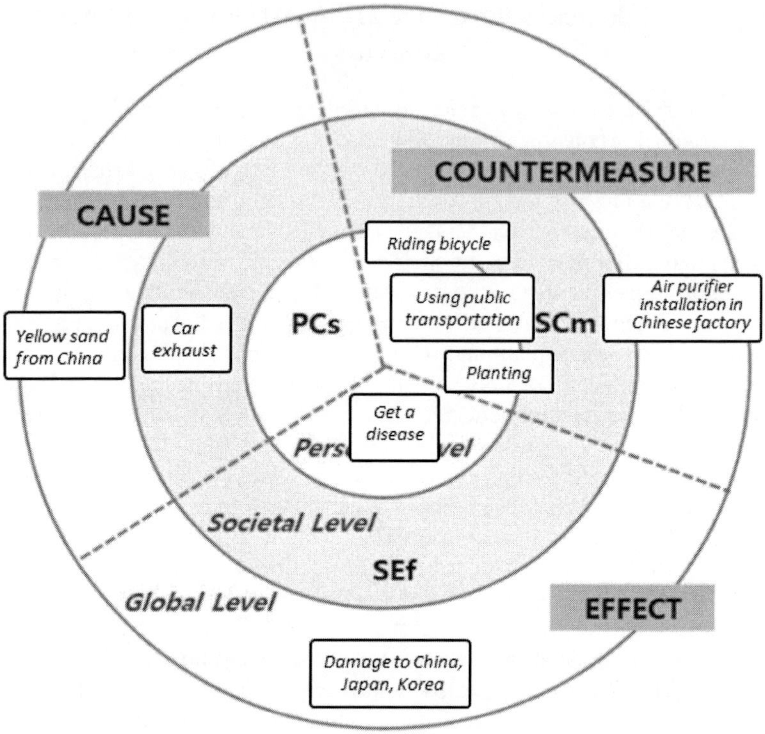

Figure 8.3: The example of pre-ICM on fine-dust (Drawing on Kim et al., 2020)

Methodological affordance 3: Assessing students' understanding of SSIs

The knowledge in SSI education includes scientific knowledge related to SSIs and the nature of science and technology. Traditionally, scientific knowledge is supposed to be hierarchical and has correct answers. Contrary to this, the nature of science and technology in SSI education is characterised by the ambivalence of science and technology, contextual dependence on individual and social-cultural characteristics, and decision-making based on moral and ethical values (Lee & Lee, 2016; Zeidler et al., 2005). The solution is problematic, and there is no correct answer. Therefore, evaluating students' understanding of SSIs is complicated and hard to investigate. As in general science education, teachers' understanding of their students helps them to organise SSI teaching and learning. Especially when there are time constraints that cover a wide range of SSI topics if these are not in the curriculum. ICMs focus on causal relationships and countermeasures rather than hierarchical relationships in traditional concept maps. ICMs also facilitate students to consider multiple contexts. However, reflecting on our previous research experience, some considerations arise when using ICMs to assess students' understanding of SSIs.

Distinguishing between idea and concept/information and knowledge

Similarly, it is sometimes difficult to distinguish whether students' answers written in ICM are ideas or concepts. Considering our climate change research as an example, students' answers in the 'Energy waste' category (232 answers) at the personal level ranged from a broad 'energy waste' to more specific 'air-conditioning overuse'. However, it is difficult to say whether students fully understand the cause of climate change from the students' answer of 'Energy waste'. They may recognise 'Energy waste' via information through climate change-related media, but may need to understand how 'Energy waste' causes climate change. Then, considering the answer 'Airconditioning overuse'–it is a more detailed answer than 'Energy waste', but it is still unclear whether students fully understand the energy production process and greenhouse gas emissions. However, if 'air-conditioning overuse' leads to additional 'increased power plant operation' or 'greenhouse gas emission', this can be classified as a clear concept.

In our fine-dust study, the effect of fine-dust on a personal level is shown in the category of 'Disease outbreak'. The responses of the pre-ICMs ranged from the broad 'may get sick' and 'disease' to specific 'sore throat', 'bronchial problems', 'lung cancer', and 'increased waste products in the lungs'. The answers to symptoms such as 'sore throat' can be ideas from students' prior experiences or information through the media. In addition, students might understand it as a respiratory disease caused by yellow dust or air pollution. Students rewrote their ICMs after receiving a community-based SSI class, and many of their post-ICM responses mentioned specific disease names. Student responses included: respiratory, brain, eye, depression, and catastrophic disabilities. They included: respiratory diseases, brain diseases, eye diseases, depression, and fatal disorders. If students understand the various types of diseases caused by fine-dust, the names of diseases can be seen as concepts. On the other hand, if students recognise the name of a disease through SSI instruction but do not understand what component of fine-dust causes the disease, it can be viewed as information.

Table 8.1: The students' answers in the 'Disease outbreak' category at the personal-level effect of fine-dust

Pre-ICMs	Post-ICMs
may get sick	serious physical condition
disease	getting sick
sore throat	headache
bronchial problems	conjunctivitis
lung cancer	itchy eyes
increased waste products in the lungs	depression
	foetal growth worse / disorder
	deformed child
	brain damage
	rhinitis
	bronchitis
	cough
	asthma
	pneumonia
	respiratory diseases

Hindering student thinking by guided ICMs

A concept map consists of concepts, connections between concepts, and linking words. However, the concept in ICMs is explained without connections and linking words since ICMs specify the occurrence and context for each cell. For example, the concept of 'Use of disposable product' in Figure 8.4. can be explained as 'Use of the disposable product is a cause of climate change at the personal level' even if there is no connection and linking word. In our climate change research, it was shown that students expanded their concepts to the connection between cause-influence-countermeasures and personal-societal-global level connection patterns for climate change. Since the connection between cause and influence cells includes '~is due to~' and the connection between influence and countermeasure cells includes '~is to solve~', the linking word can be omitted. In contrast, personal-societal-global level connections may not be sequential causation. In Figure 8.4, the concepts are connected by arrows pointing from the inside to the outside of the ICM (e.g., Decreasing farmland -> Desertification -> Global warming). However, in (1) Heatstroke -> Heat wave, the direction of the arrow should be changed from the outside to the inside of the ICM because societal consequences 'Heat waves' affect individual health 'Heatstroke'. (2) Flood sufferer damage -> Floods is also the same. By including linking words in ICMs, students tend to draw connections in similar patterns, counter-clockwise (cause -> influence -> countermeasure) and outward from the centre of the ICM (personal -> societal -> global levels), without thinking carefully about relationships between concepts.

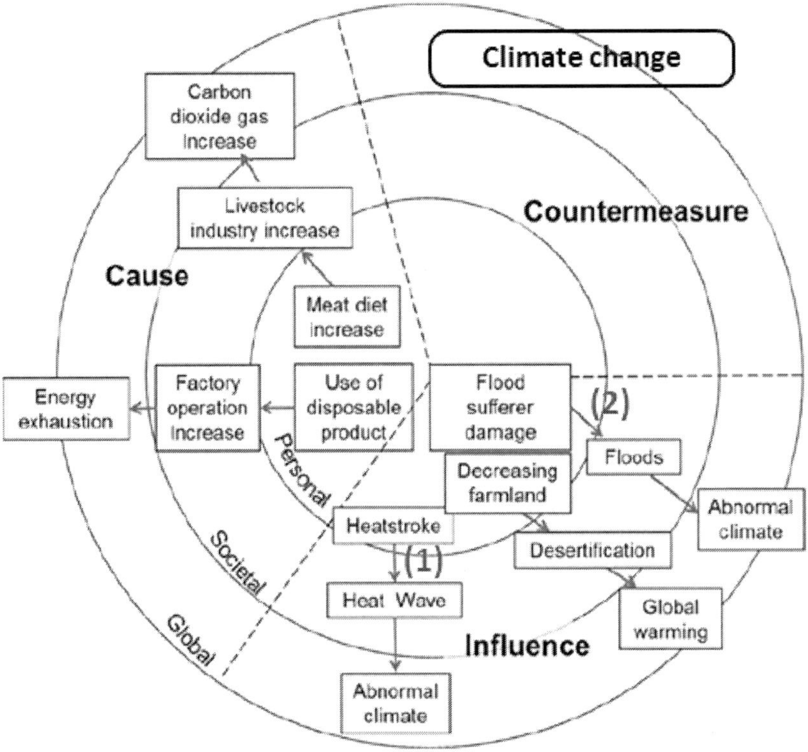

Figure 8.4: Personal-societal-global connections (Drawing on Mun et al., 2014)

In Table 8.2, the connection patterns between ideas in our fine-dust study are mostly no connection or only cause-influence-countermeasure connection in the same context in pre-ICMs. In post-ICM, most connections are in the same context or occurrence, whereas connections between different contexts and occurrences are rare. Because concepts in ICMs facilitate extending concepts to cause-effect-countermeasures and personal-societal-global levels, students unconsciously connect concepts primarily in a direction that diverges into a counter-clockwise, cause-effect-countermeasure, or personal-societal-global context, according to the ICM framework. In addition, students can think along the boundaries between the ICM cells rather than expanding their thoughts freely like a traditional concept map. Therefore, students may need to consider the connection between concepts while crossing the boundary between contexts and occurrences.

Table 8.2: Changes of connections between concepts (Kim et al., 2020)

Case	Pre-ICMs	Post-ICMs
1	None	None
2	None	C-E-Cm in the same context
3	None	C-E-Cm in the same context P-S-G in the same occurrence P-S-G in the different occurrences
4	C-E-Cm in the same context	C-E-Cm in the same context
5	C-E-Cm in the same context	C-E-Cm in the same context P-S-G in the same occurrence
6	C-E-Cm in the same context	C-E-Cm in the same context P-S-G in the same occurrence P-S-G in the different occurrences Connections within a cell

* C: Cause, E: Effect, Cm: Countermeasure, P: Personal, S: Societal, G: Global

In addition, the student's response to the countermeasure becomes the endpoint of the connection. That is, students' thinking stops without consideration for countermeasures to resolve the issue. Sometimes the measures taken to resolve an issue create other problems that affect individuals and societies. For example, 'Air purifier installation in Chinese factory' in Figure 8.3 can be a solution to fine-dust, while this countermeasure can affect the Korean economy, China's economy and society, and generate friction between countries. A possible scenario for how this measure could affect the Korean economy is that the installation of an air purifier in a

Chinese factory could cause wage cuts, workforce cuts, and factory closure due to the economic burden on the Chinese factory. Then, the critical question to consider here is, 'How many international companies are there in China?', 'How many Koreans work in this factory?', 'If a Chinese factory goes out of business, will it affect the price of Korean companies' products?'. Teachers can help students think critically about countermeasures by prompting students about their potential economic, religious, social, and international impacts when their countermeasures in ICMs do not show connections to other concepts. Moreover, we suggest that the cell boundaries of the ICMs be drawn as dotted lines or faded to relieve the constraints imposed by the frame of the ICMs.

Identifying reasoning for changes of pre-concepts

In our study on changes in students' understanding of fine-dust, students were instructed to write a pre-ICM and then overlap the post-ICM with a different colour pen at the end of the SSI lesson. When writing the post-ICM, students removed or moved concepts in pre-ICM. For example, in Figure 8.5, 'Yellow sand from China', written as a global cause of fine-dust, was deleted in the post-ICM process. Teachers can expect that students' understanding of the cause of fine-dust has improved since they have eliminated the misconception of fine-dust 'yellow sand'. However, it is difficult to know whether the students understood the difference between fine-dust and yellow dust and removed it or whether they recognised and removed 'yellow sand' as not the cause of fine-dust. In addition, 'Car exhaust' changed from a societal to a personal level. Teachers can recognise that students perceive 'Car exhaust' as the cause of fine-dust at the personal level rather than the societal level.

On the other hand, teachers cannot determine why students view 'Car exhaust' as a cause at the societal level beforehand and whether the reason for changing it to the personal level is based on evidence or personal experience. Teachers can comprehend the context and occurrence from the position of the changed pre-concept, but it is difficult to grasp the specific reasoning for the students' concept change. To compensate for this, we suggest adding a reasoning section, 'Why do you change your idea?' on ICM, as shown in Figure 8.6. For example, if the pre-idea A of SSI is changed from a societal level effect to a societal level cause, students write the change path and specific reasons for idea A in the reasoning section why idea A was changed.

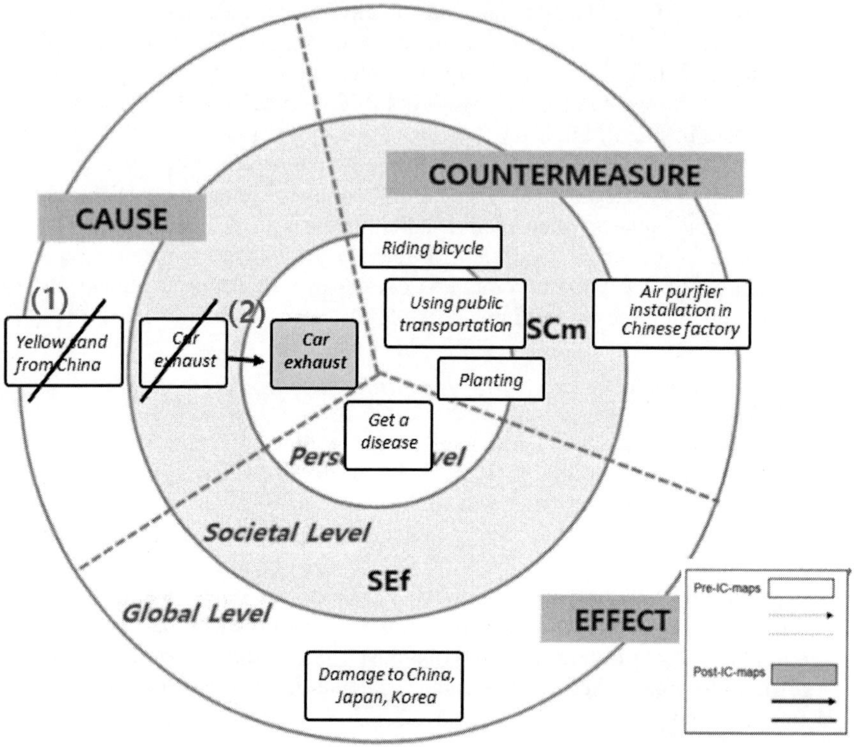

Figure 8.5: Example of pre-and post-concept change on fine-dust (Drawing on Kim et al., 2020)

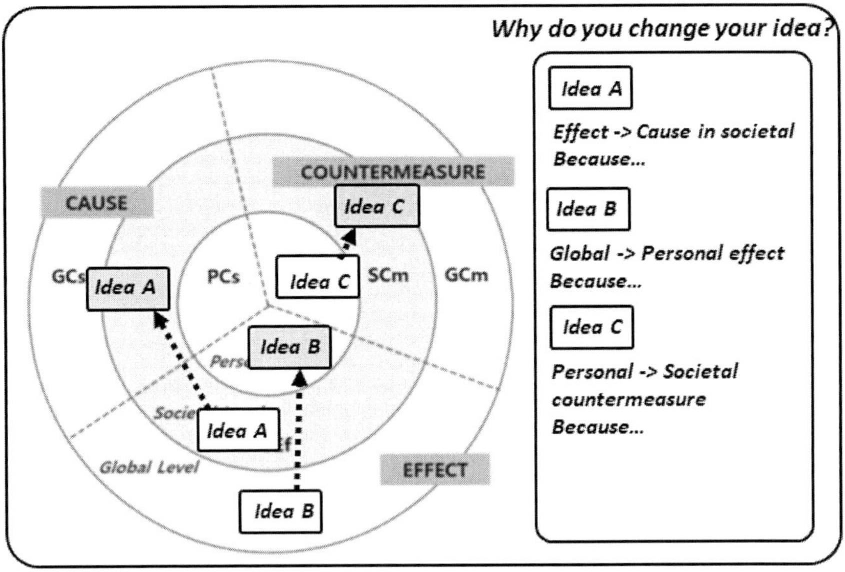

Figure 8.6: The ICMs for understanding students' reasoning

Conclusion

SSI education aims to cultivate citizens who can engage in rational decision-making and practice action on SSIs caused by science and technology. Identifying students' understandings of SSIs is essential if we are to achieve this goal. This chapter explores a design-based research cycle by drawing on our previous research using the ICMs which we developed to investigate students' understandings of SSIs. By reflecting on our research, we identify the methodological affordances and considerations of ICMs in SSI instruction and propose teaching and learning strategies. The methodological affordances of ICMs are selecting SSI topics, constructing SSI lesson content, and assessing students' conceptual understandings of SSIs.

It is difficult for researchers to fully implement theory-based instructional tools in real-world contexts. However, the design-based research approach allows researchers to identify new problems and unexpected variables through iterative implementation in authentic classrooms rather than laboratories (Collins et al., 2004). This process allows researchers to bridge the gap between theory and practice to ensure optimal learning outcomes. In addition, when reanalysing our previous studies through the design-based research method, we may be exposed to various biases during the analysis process (Anderson & Shattuck, 2012; Barab & Squire, 2004) because complicated analyses of all previous data, reflections of researchers and teachers, and comparison of two studies are required. In order to resolve this issue, researchers who fully understood two previous studies participated in the analysis and opinions were collected through several discussions.

Depending on the social and cultural context, SSI topics can be broad or specific for participants (Hancock et al., 2019; Jang et al., 2012). To select SSI topics, teachers can use ICMs to identify the sensitivity and relevance of students' issues. In addition, through pre-ICMs, teachers can visually grasp students' understanding of SSIs and identify misconceptions. Based on this, ICMs can be used to configure learning goals and the content of SSI lessons. Furthermore, there are several considerations when applying ICMs to assess students' understanding of SSIs: 1) The setting of evaluation criteria for the answers to ICMs written by students may need to be clarified. Because SSI education must accompany teaching on causes, influences, and countermeasures related to SSIs (Choi et al., 2012; Dove, 1996; Lee at al., 2012; Morgan & Moran, 1995), student answers cannot be accurately made with concepts and knowledge only; 2) The guidelines of ICMs can promote the expansion of students' thinking while at the same time limiting their thinking within ICMs; 3) It was possible to recognise changes in students' understanding through pre-ICMs, however it was challenging to grasp students' reasoning about concept changes. For using ICMs in SSI education, there are several suggestions.

First, teachers should give students an opportunity to practise writing ICMs before this activity. In our previous study, for most of the students, this was their first time using ICMs, and they wrote responses in personal-level cells only or wrote concepts or ideas that were not clear. Therefore, it is important to familiarise students with the need to write ideas or concepts in multiple cells when writing an ICM.

Second, when using ICMs, teachers should plan lessons to ask students to consider the economic, cultural, and moral perspectives of SSI. ICMs help explore various contexts and occurrences. The answers mainly focus on scientific and environmental aspects, although SSI education emphasises the importance of multiple perspectives. Teachers should help students consider different perspectives related to SSI.

Third, groups engaging with ICMs should comprise students from a variety of backgrounds and perspectives. SSI topics may involve different approaches and understandings depending on learners' family backgrounds, prior experiences, religious beliefs, and cultural and social backgrounds. Teachers can investigate students' previous understandings of SSIs through individual ICM activities and then form groups composed of students with various understanding patterns, in order to stimulate rich discussions.

Last, when researchers plan to develop and implement SSI lessons, they need to consider the design-based research methodology to help teachers adapt. SSI education inherently contains complexity and uncertainty and needs to be modified according to various contexts and situations. These characteristics of SSI can cause difficulties for teachers in SSI teaching. Therefore, researchers maintain a long-term cooperative relationship with teachers through the design-based research methodology and need to conduct repeated testing and redesigning so that teachers can adapt to unfamiliar SSI lessons in actual classes.

Implications in SSI education

ICMs have educational significance in that they play the role of a facilitator so that students can consider SSIs in various contexts and occurrences. There is a benefit in that teachers and researchers can visually grasp students' understandings of SSI. In addition, since teachers can investigate students' prior knowledge and awareness of issues with ICMs and use them in selecting topics and constructing content in SSI lesson plans, this can be expected to address difficulties in preparing for SSI lessons. Furthermore, the design-based research methodology contributed to us reflectively examining our previous studies and deriving practical strategies for implementing a new SSI lesson using ICMs. This study presented methodological affordances and considerations when applying ICMs to SSI classes. However, since this study was conducted with Korean students, there are limitations in applying it to Australia and other countries with different geographical, contextual, and cultural characteristics. Therefore, researchers need to apply ICMs to teachers and students in each country and explore concept attributes. Finally, researchers need to plan teacher workshops and distribute guidebooks to support teachers to use ICMs in SSI classes.

References

Andersson, B., & Wallin, A. (2000). Students' understanding of the greenhouse effect, the societal consequences of reducing CO2 emissions and the problem of ozone layer depletion. *Journal of Research in Science Teaching: The Official Journal of the National Association for Research in Science Teaching*, *37*(10), 1096-1111.

Anderson, T., & Shattuck, J. (2012). Design-based research: A decade of progress in education research? *Educational Researcher*, *41*(1), 16-25.

Barab, S., & Squire, K. (2004). Design-based research: Putting a stake in the ground. *The Journal of the Learning Sciences, 13*(1), 1-14.

Boon, H. J. (2010). Climate change? Who knows? A comparison of secondary students and pre- service teachers. *Australian Journal of Teacher Education, 35*(1), 104-120.

Boon, H. J. (2014). Teachers and the communication of climate change science: A critical partnership in Australia. *Procedia-Social and Behavioral Sciences*, *116*, 1006-1010.

Boon, H. J. (2016). Pre-service teachers and climate change: A stalemate? *Australian Journal of Teacher Education (Online)*, *41*(4), 39-63.

Boyes, E., & Stanisstreet, M. (1997). The environmental impact of cars: Children's ideas and reasoning. *Environmental Education Research, 3*(3), 269-282.

Boylan, C. (2008). Exploring elementary students' understanding of energy and climate change. *International Electronic Journal of Elementary Education*, *1*(1), 1-15.

Brown, A. L. (1992). Design experiments: Theoretical and methodological challenges in creating complex interventions in classroom settings. *The Journal of the Learning Sciences*, *2*(2), 141-178.

Choi, K., Lee, H., Shin, N., Kim, S., & Krajcik, J. (2012). Re-conceptualization of scientific literacy in South Korea for the 21st century. *Journal of Research in Science Teaching, 48*(6), 670-697.

Collins, A. (1992). Toward a design science of education. In E. Scanlon & T. O'Shea (Eds.), *New directions in educational technology*. Berlin: Springer-Verlag.

Collins, A., Joseph, D., & Bielaczyc, K. (2004). Design research: Theoretical and methodological issues. *The Journal of the Learning Sciences, 13*(1), 15-42.

Dawson, V. (2015). Western Australian high school students' understandings about the socioscientific issue of climate change. *International Journal of Science Education*, *37*(7), 1024-1043.

Dove, J. (1996). Student teacher understanding of the greenhouse effect, ozone layer depletion and acid rain. *Environmental Education Research, 2,* 89-100.

Fowler, S. R., Zeidler, D. L., & Sadler, T. D. (2009). Moral sensitivity in the context of socioscientific issues in high school science students. *International Journal of Science Education*, *31*(2), 279-296.

Hancock, T. S., Friedrichsen, P. J., Kinslow, A. T., & Sadler, T. D. (2019). Selecting socio-scientific issues for teaching: A grounded theory study of how science teachers collaboratively design SSI-based curricula. *Science & Education*, *28*, 639-667.

Jang, J., Mun, J., Ryu, H. S., Choi, K., Joseph, K., & Kim, S. W. (2012). Korean middle school students' perceptions as global citizens of socioscientific issues. *Journal of the Korean Association for Science Education*, *32*(7), 1124-1138.

Karpudewan, M., Roth, W. M., & Abdullah, M. N. S. B. (2015). Enhancing primary school students' knowledge about global warming and environmental attitude using climate change activities. *International Journal of Science Education*, *37*(1), 31-54.

Kim, G. (2016). The effects of community-based socioscientific issues programs on promoting students' understanding of community issues and character and values as citizens (unpublished doctoral dissertation). Ewha Womans University, Korea.

Kim, G., Ko, Y., & Lee, H. (2020). The effects of community-based socioscientific issues program (SSI-COMM) on promoting students' sense of place and character as citizens. *International Journal of Science and Mathematics Education*, *18*, 399-418.

Kim, G., & Lee, H. (2017). Effects of community-based SSI programs on promoting middle school students' understanding of issues and character and values as citizens: Focused on fine dust issues. *Journal of the Korean Association for Science Education*, *37*(6), 911-920.

Kim, G., & Lee, H. (2021). A case study of community-based socioscientific issue program: Focusing on the abandoned animal issue. *Journal of Biological Education*, *55*(4), 380-394.

Kim, G., Mun, K., & Lee, H. (2020). Exploration of middle school students' ideas of fine dust issues using issue concept maps. *Asia-Pacific Science Education, 1*, 1-12.

Kolstø, S. D. (2001). Scientific literacy for citizenship: Tools for dealing with the science dimension of controversial socioscientific issues. *Science Education*, *85*(3), 291-310.

Lee, H., Chang, H., Choi, K., Kim, S., & Zeidler, D. L. (2012). Developing character and values for global citizens: Analysis of pre-service science teachers' moral reasoning on socioscientific issues. *International Journal of Science Education, 34*(6), 925-953.

Lee, H., & Lee, H. (2016). Contextualized nature of technology in socioscientific issues. *Journal of the Korean Association for Science Education, 36*(2), 303-315.

Lee, H., Yoo, J., Choi, K., Kim, S. W., Krajcik, J., Herman, B. C., & Zeidler, D. L. (2013). Socioscientific issues as a vehicle for promoting character and values for global citizens. *International Journal of Science Education, 35*(12), 2079-2113.

Morgan, M. D., & Moran, J. M. (1995). Understanding the greenhouse effect and the ozone shield: An index of scientific literacy among university students. *Bulletin of the American Meteorological Society, 76*(7), 1185-1190.

Mueller, M. P., & Zeidler, D. L. (2010). Moral–ethical character and science education: Ecojustice ethics through socioscientific issues (SSI). *Cultural studies and environmentalism: The confluence of ecojustice, place-based (science) education, and indigenous knowledge systems* (pp. 105-128).

Mun, K., Kim, G., Kim, S. W., & Krajcik, J. (2014). Exploration of high school students' concepts about climate change through the use of an issue concept map (IC-Map). *In International Conference on Science Education 2012 Proceedings,* 209-222. Springer Berlin Heidelberg.

Sadler, T. D., Barab, S. A., & Scott, B. (2007). What do students gain by engaging in socioscientific inquiry? *Research in Science Education, 37*, 371-391.

Sadler, T. D., & Donnelly, L. A. (2006). Socioscientific argumentation: The effects of content knowledge and morality. *International Journal of Science Education, 28*(12), 1463-1488.

Sadler, T. D., & Zeidler, D. L. (2005). The significance of content knowledge for informal reasoning regarding socioscientific issues: Applying genetics knowledge to genetic engineering issues. *Science Education, 89*(1), 71-93.

Sandoval, W. A., & Bell, P. (2004). Design-based research methods for studying learning in context: Introduction. *Educational Psychologist, 39*(4), 199-201.

Zeidler, D. L., & Keefer, M. (2003). The role of moral reasoning and the status of socioscientific issues in science education: Philosophical, psychological and pedagogical considerations. *The role of moral reasoning on socioscientific issues and discourse in science education* (pp. 7-38).

Zeidler, D. L., & Sadler, T. D. (2007). The role of moral reasoning in argumentation: Conscience, character, and care. *Argumentation in science education: Perspectives from classroom-based research* (pp. 201-216).

Zeidler, D. L., Sadler, T. D., Applebaum, S., & Callahan, B. E. (2009). Advancing reflective judgement through socioscientific issues. *Journal of Research in Science Teaching: The Official Journal of the National Association for Research in Science Teaching*, *46*(1), 74-101.

Zeidler, D. L., Sadler, T. D., Simmons, M. L., & Howes, E. V. (2005). Beyond STS: A research-based framework for socioscientific issues education. *Science Education*, *89*(3), 357-377.

Zeidler, D. L., Walker, K. A., Ackett, W. A., & Simmons, M. L. (2002). Tangled up in views: Beliefs in the nature of science and responses to socioscientific dilemmas. *Science Education*, *86*(3), 343-367.

Gahyoung Kim (BScE, MPhys, PhD) - Monash University, School of Curriculum Teaching and Inclusive Education, Wellington Rd, Clayton VIC 3800, Australia.

Gahyoung Kim is an educational researcher interested in designing and implementing community-based education in formal and informal settings on Socioscientific Issues. Especially, engaging students in learning and applying their knowledge in everyday life as an activist in their community. Gahyoung has diverse teaching experience for pre-service teachers and informal educators. Also, her research includes quantities and qualitative methodologies.
https://orcid.org/0000-0002-3583-2101

Kongju Mun (BEd, PhD) - Dongduk Women's University, Liberal Arts College, 60 Hwarang-ro 13-gil, Seongbuk-gu, Seoul, 02748, Korea.

Kongju Mun is an Assistant Professor in the Liberal Arts College at Dongduk Women's University. She is developing a curriculum that integrates science and humanities for non-science major students, and pioneering a novel form of science education for female college students. Her research interests include the convergence of science and art, out-of-school science education, and SSI education.
https://orcid.org/0000-0003-4494-3871

CHAPTER 9

USING NETNOGRAPHY TO SURFACE 'SUCCESS STORIES' FROM OUT-OF-FIELD TEACHERS OF STEM

MARGARET JAKOVAC

Abstract

Some six decades after Conant (1963) gave it a name, the out-of-field teaching phenomenon is a "widespread and global epidemic in education" (Du Plessis, 2020, pp. 76-77). Schools misassign teachers outside their qualification area (Hobbs & Törner, 2019) to address the growing teacher shortage and patchy workforce data. Public discourse and scholarly research tend to foreground a deficit perspective, with 'success' stories of out-of-field teachers largely missing from both discourses. Through the theoretical lens of identity and practices, this research aims to explore Facebook groups in which out-of-field science, technology, engineering, and mathematics (STEM) high-school teachers narrate experiences of success over time. The researcher will be a participant observer because of her teaching background. Up to 10 out-of-field teacher members of Facebook groups will be interviewed and asked to draw what contributed to their experiences of success. I will ethically collect data using netnography (Kozinets, 2020), a form of ethnography tailored to social media, and then analyse using Reflexive Thematic Analysis (Braun & Clarke, 2006). The research is methodologically novel to explore the out-of-field-teaching phenomenon, particularly to find a "collective professional voice" for those whose voices education undervalues (Thomson & Riddle, 2018, p. 126).

Key words: Out-of-field teaching, Netnography, Success stories, STEM

I am an occasional casual, relief and temporary high school teacher, though primary-school trained. A taxonomy of teachers would also categorise me as an out-of-field teacher. The latter is how I largely identify when I work in education. Often, I am assigned to teach science, technology, engineering and mathematics (STEM) classes, so my shingle for schools has become 'out-of-field-STEM teacher'. But my identity is further hybridised – as journalist and content marketing writer with niches in education, business and technology. I draw on communications and education disciplines to explore 'success' stories about the out-of-field teaching phenomenon.

The out-of-field teaching phenomenon: Its context

Due to the pervasive teacher shortage across Australia, there is an exponential increase in those "teaching a subject without a background in the discipline or preparation for teaching it" (Vale et al., 2022, p. 3), i.e., being out-of-field. It might appear as if the declining teacher workforce has singularly caused out-of-field teaching, but the out-of-field teaching phenomenon is a "complex problem" (Hobbs et al., 2022, p. 8). Indeed, scholarly research lacks consensus on a definition (Hobbs et al., 2022, p. 11) and, according to Rochette, the phenomenon as a "deficit or opportunity" (2022, p. 101). Interrogating the 'opportunity' means requesting whom, what, in which context(s), and to where. A possibility is the opportunity for a teacher to eventually acquire an in-field designation, shucking off the out-of-field label. This would mean the apex of a successful out-of-field teacher is indeed one who has 'escaped' this subset of teachers. This positioning aligns with a leading scholar's view that out-of-fieldness does not always equate to being "ill-equipped to be an effective teacher" (Hobbs, 2013, p. 293), therefore a reframing to 'strengths based' (He, 2019) may be in order. My contribution and methodological path seek to focus on the 'success' narratives of out-of-field high-school teachers of STEM, therefore the 'opportunities' for teachers, in particular contexts. These include geographical (inner-city, metropolitan, outer-metropolitan, regional, rural, remote, and very remote), by sector (government, Catholic, or independently run schools), and the socio-economic status of parents/guardians (from extremely disadvantaged to extremely advantaged) and based on the National Assessment Program-Literacy and Numeracy (NAPLAN; ACARA, 2020).

Returning to the workforce shortage in school education, it is one reason teachers are asked or directed to teach subjects or stages outside of their qualification areas. The situation is expected to worsen, particularly in

Australia's most populous state of NSW, with an audit office report citing increased student numbers, an ageing teacher workforce and shrinking pre-service teacher numbers (Audit Office, 2019, p. 1). Internal education department documents have forecast NSW will "run out of [new public school] teachers by 2026" (Baker, 2021, para. 1). That decline will supposedly occur even though that department was "not accurately tracking the supply and demand … [of high school science, technology, engineering and mathematics (STEM) teachers] due to incomplete data" (Audit Office, 2019, p. 1). The NSW department's focus on out-of-field STEM teaching is also echoed in public discourse and scholarly research, despite other subject areas also experiencing shortages of qualified teachers (Weldon, 2016; Caldis, 2017; AITSL, 2018). Problematically, though, patchy teacher workforce data is reflected across the country and globe (Department of Foreign Affairs and Trade (DFAT), 2019; Du Plessis, 2020; Hobbs & Törner, 2019; O'Connor & Thomas, 2019; Weldon, 2016; Weldon, 2018; Vale, 2019). A 2022 national summit on out-of-field teaching has also recommended consensus on the phenomenon's definition and a "systematic and system-wide data" collection (Hobbs et al., 2022, p. 6). While the problem cannot yet be comprehensively quantified, misassigning teachers (Quirke, 2022) is a quick fix to a "systemic" phenomenon (Vale & Drake, 2019, p. 195), that is a "widespread and global epidemic in education" (Du Plessis, 2020, pp. 76-77). Implicit here is a deficit discourse about the impact of out-of-field teaching, which requires unpacking.

Much education research about the out-of-field teaching phenomenon has highlighted its negative impacts over the six decades since Conant's 1963 seminal monograph raised the significant "misuse of teachers" in the US (as cited in Ingersoll, 1999, p. 26). The deficit discourse has predominated in research internationally through various analytical lenses, including teachers' expertise and lived experience (Vale & Drake, 2019, p. 211; Quirke, 2022; Ingersoll, 1999; Rochette, 2019); their identity, confidence, competence (Du Plessis, 2020, p. 280; Hobbs et al., 2020, p. 38; Caldis, 2017; Nadelson et al., 2013; Shernoff et al., 2017), and impacts on student learning outcomes (Du Plessis, 2020; Porsch & Whannell, 2019), amongst others. Regarding public discourse, Australian media representations of teachers and teaching – not just out-of-field – highlight quality as declining, as Mockler's (2022) analysis of over 65,000 media articles from 1996 to 2020 showed. My prior research into online Australian media representations of out-of-field teaching of STEM offers a smaller snapshot, being based on 26 curated articles spanning five years to December 2021. The analysis suggested the media problematised out-of-field STEM teaching as indicative of the teacher shortage, contributing to teacher overwork and negatively impacting student learning outcomes through shortcomings in teaching quality and teacher quality (Jakovac, 2022). STEM is oft-mentioned in public discourse, but also in research into the out-of-field teaching phenomenon.

Despite policymakers, governments, and school education systems globally prioritising it, STEM education is a contentious research field (Milner-Bolotin, 2018). Mohamad Hasim et al. (2022, p. 1) posit that STEM is a "crucial global driver" to developing "various aspects of modern society, such as the economy, technology, education, and skills of the 21st century workforce". Meanwhile, Kurup et al. (2021, p. 1121) describe STEM as an "innovative ... and key education practice" that offers a path for inquiry learning and creativity. They continue that it becomes powerful when taught in an "integrated and interdisciplinary" way (p. 1121), rather than combining subjects under four disciplines. However, at classroom level, there is much diversity and confusion regarding how teachers can and should implement and integrate STEM (Siegel & Giamellaro, 2022). Education systems across the world inconsistently integrate STEM into K-12 classrooms (Kayan-Fadlelmula, 2022), making it challenging to compare pedagogical best practices and modify them for specific contexts. Others offer simpler, overarching guides, such as Williams (2019, p. 1) who argues that integrating STEM education requires "no new pedagogies", instead, teachers should follow STEM principles. Amongst this quandary is a "global shift in educational paradigms" stressing STEM's importance (Kayan-Fadlelmula et al., 2022. p. 1). Concerns about the long-term economic impacts of out-of-field teaching of STEM subjects also feature strongly in Australia (Finkel, 2020), with education ministers agreeing in 2016 to a decade-long national STEM school education strategy (Australian Government Department of Education, 2023, para. 1). Following on from this move, Federal Australian policymakers had recommended out-of-field STEM teaching be phased out in the five years to June 2022 (Parliament of Australia, 2017, p. xviii), but state and territory governments neither adopted nor actioned it (D. Laming, 2022). Hence, there is a deep societal push for quality STEM teaching in schools, underpinned by an assumption that out-of-field teachers of STEM are not an 'ideal' part of this vision.

Yet, all teachers must abide by the Australian Professional Standards for Teachers (AITSL, n.d.), the national yardstick for quality. The standards include one requiring content knowledge and pedagogical knowledge, without specifying the knowledge for a specific subject be gained through a university teaching qualification or classroom experience (AITSL, n.d.). Australian education jurisdictions, however, have varying registration requirements for which subjects and stages teachers can teach, and whether years of classroom experience can convert a teacher's out-of-field area to in field. The NSW teacher registration authority recognises such experience (two years in the past five) for teaching subjects other than English, science, and mathematics (Baker, 2022, para. 11). Elsewhere, that authority's policy states it will not accept classroom experience in "lieu of academic study for teaching subjects" (NESA, 2021, p 1). Valuing teaching experience aligns with Du Plessis' (2020) finding that some teachers she had interviewed felt 'in field' within three to five years of teaching a subject out of field (p. 24). At the crux is an appreciation that teachers can build expertise over time, such as along a continuum of routine to classic expertise (Hatona & Inagaki, 1986). But public discourse rarely reflects expertise building regarding out-of-field teaching, and only in a limited way in scholarly research. The latter includes insights through the lens of a sense of teacher agency (Du Plessis, 2020), resourcefulness (Beswick et al., 2016), positive identity, passion, or pursuing an interest in a subject (Hobbs, 2013; Pacana et al., 2019).

Surfacing silenced stories of STEM teaching success

My research seeks to ethically 'surface' success stories about out-of-field STEM teachers working in Australian high schools over time, that is not at a singular point. Casting the net wide also acknowledges teachers' journeys to success may be varied. I also harness a broad definition of out-of-field teaching of STEM because this may ensure I can find participants. Such teachers may be required to teach a discrete subject that falls into a STEM discipline (such as biology, science, geology, chemistry, physics), combined/integrated or interdisciplinary STEM. Admittedly, including the last two categories may increase my chances of finding sufficient out-of-field teachers because few have qualifications in all disciplines. A few Australian universities offer post-initial-teacher education qualifications in STEM.

Developing, refining and maintaining content knowledge, and pedagogical content knowledge (Shulman, 1986) for each STEM discipline *and* the same for integrating those disciplines is "unrealistic" for teachers (Siegel et al., 2022, p. 580). The Australian Curriculum (n.d.) suggests rather than directs teachers to achieve STEM interdisciplinarity. As mentioned above, researchers lack consensus about how to integrate STEM education well. Therefore, out-of-field teachers of STEM may lack access to rubrics, best practices, mentorship or understandings about exemplary integrated STEM teaching practices. This study does not seek to rate or judge their teaching practices, rather explores their narrations and interpretations of successful STEM teaching. Hence, the scope for this research is to collect data from such teachers narrating their perceptions of success in the full suite of ways in which STEM is taught in their schools.

Starting with a pastiche: Building a conceptual framework

I draw on various sociological and sociocultural theories to sketch out a conceptual framework with teacher identity and practices as my foci. To harness identity as a lens of analysis, I ruminated on the contentious array of research espousing sociological, sociocultural or psychological theoretical frames (Beijaard et al., 2004). Darragh's (2016, p. 29) critique of identity research in mathematics education posits that psychological theories would be apt if a researcher regards 'success' as an acquisition. For an out-of-field teacher of STEM, that might be akin to conducting one successful class and considering oneself 'successful' from then on, which I argue would be unlikely. Also, this stance does not align with the requirement for teachers to maintain a continuous professional development cycle (AITSL, 2022). Nor does it appreciate the numerous contexts in which teaching occurs, meaning one lesson plan can achieve different results depending on various factors, including the student cohort, type of school, and resources available to the teacher.

Meanwhile, Darragh's review of sociological and socio-cultural theories finds they suggest that identity is performative and action-based within a social context or group (2016, p. 2). Out-of-field teaching research scholar Du Plessis offers a possibly useful theory in her Context-Consciousness Understanding Development (C-CUD) Theory (2020). It draws on social-constructivism and other theories and philosophies to view the phenomenon through the perspectives of out-of-field teachers' lived experience and their school leaders. It recognises such teachers' experience and the "social, political and historical contexts" (Du Plessis & McDonagh, 2021, p. 5) in which they occur as a "dialogic, practical and situated action" (Du Plessis & McDonagh, 2021, p. 4). I entertained C-CUD theory because of its sensitivity to contexts and their dynamism, but my ambit does not include school leaders, so it is not appropriate. In appreciation of out-of-field teachers' success as not solely an internalised disposition, I therefore choose sociocultural theories which ascribe to Wenger's sociocultural elaboration of identity as a "constant becoming" rather than an object (1998, pp. 153-154). Therefore, I am influenced by the work of two other out-of-field teaching scholars who have employed boundary crossing (Hobbs et al.,

2019), and positioning theory (Rochette, 2022). They afford me potential insights to scope the interplay between identity, contexts, and building expertise over time. Another theory from education is useful, too: that of practice architectures. I will briefly explain the triad of theories below.

Boundary Crossing Model

Hobbs (2013) adapted the term 'boundary crossing' as a metaphor for out-of-field teaching to describe the "*spatial* boundary crossing between familiar and unfamiliar fields … [and the] *experiential* and *temporal*" (Hobbs, 2020, p. 727). So, as an English-qualified teacher is assigned to a science class, then perhaps to a humanities class, they may move between the unfamiliar (out-of-field) to the vaguely familiar (near-field), then back to their English teaching (in-field). Such boundary crossing can happen within a teaching day, and as mentioned with the teacher shortage, is commonplace within contemporary teaching careers in Australia. The Boundary Between Fields Model (Hobbs, 2013) is pertinent to my research because it recognises teachers may need to have a somewhat adaptable identity as they operate in, near, and/or out of field contexts. The model also features an adaptability scale for teaching out-of-field, i.e., allowing teachers to describe themselves as "just filling in, making the most of it, or pursuing an interest" (Hobbs, 2013, p. 291). My study hones in on the latter element. As for the resources out-of-field teachers use to navigate subject boundaries in their work, the model also provides a useful term: "boundary objects" (Hobbs et al., 2019, p. 105). Using the Boundary Crossing Model, I have methodological tools to explore out-of-field teacher identity, their dimensions of success over time, and across subject 'borders'.

Positioning theory

Positioning theory offers a useful lens to explore teacher identity in their work contexts. Rochette was innovative in adapting Clandinin and Connelly's (1996) school staff "stories as a framework within the positioning triad" to out-of-field teaching (Rochette, 2022, p. 110). Clandinin and Connelly (1996, in Rochette, 2022, p. 110) described three types of stories: "cover", "sacred", and "secret". These stories acknowledge a "mis-/match between what is said and done" (Hobbs et al., 2019, p. 90), shedding light on teacher agency. For example, a school administrator would espouse a 'cover' story of STEM being taught well, whereas "teachers, policymakers, theoreticians" might differ in their sacred stories which see STEM practice as theory driven (Rochette, 2022, p. 110). At the classroom practice level, teachers could have a different perspective on their "secret" stories that diverge from the cover and sacred stories (Rochette, 2019, p. 98). I am particularly interested in exploring out-of-field STEM teachers' narratives about success and their understanding of any variance from those of school leaders', policymakers, scholars, and others. That variance may shed some light on what Brown et al. (2022, p. 295) denote as an under-researched aspect of out-of-field teaching, i.e., their overconfidence.

Practice architectures

My research delves into teachers' practices as well, which is why practice theory (Bourdieu, 1977) underpins my framework. Drawing on anthropology and sociology, practice theory explores how society, culture, and social life depend upon and develop through practice and practices. Bourdieu describes practice as that which implies cognition and construction, creating "(taxonomies) which organize perception and structure practice" (1977, p. 97). In other words, practices establish and confirm meaning. Researchers embracing this theory see the social world as "open, contingent, transitory and horizontal" (Spaargaren et al., 2016, p. 6), finding a "middle ground" between individual actions, values and choices compared with societal structures. That complexity parallels the diversity, fluidity, and reality of out-of-field teaching. Kemmis (2022) offers a more nuanced perspective on the field of education. His theory acknowledges that teachers' practices are the products of others' decisions and actions and that these shape practice architectures (Kemmis, 2022, p. 1). This framing casts the scene as a tussle between teacher agency and other actors in the contexts in which they work and the multiple factors at play. I am interested in what has contributed to out-of-field teachers' success from their own perspectives.

Right-sizing an ethnographic approach

Focusing on teachers' identities and practices can be enriching if I take heed of their context, which steers me towards ethnography, offering many theoretical and methodological approaches for data collection and analysis. Kendall and Thangara (2013, as cited in Trainor & Grau, p. 82) highlight my dilemma of choice regarding an ethnographical approach due to "the … complexity resulting from the expanded diversity of ethnographic disciplines, fields, epistemologies, ideologies, and theoretical lenses over the past century; and the explosion of analytic and presentational approaches over the past forty years". An additional challenge is that education research has a "well-known diversity in assumptions, theories, objectives and methods"

(Lund et al., 2020, p. 1). This is due to education not being a single discipline, but multidisciplinary (Lund et al., 2020, p. 15), and suffering an apparent "dispersion of discourse and an unstable epistemic community" (Lund et al., 2020, p. 16). The lack of cohesion, and my background in communications, prompts me to investigate a methodological approach, netnography (Kozinets, 2020), which is outside the field of education, to explore a phenomenon in education. Therefore, to understand social media, I see value in selecting a theory from the discipline of communications. This is in contrast to most research into media representations of teachers (Baroutsis, 2019; Mockler, 2022) which uses education as its starting point. Netnography also aligns well with my aim of doing ethnographic research. Therefore, my study will be significant and novel methodologically because there is little research using netnography to explore out-of-field teaching, those who teach all the STEM disciplines, or out-of-field teaching of mathematics or science (as a component of STEM).

My research question:

How and why do Australian out-of-field high-school teachers of STEM portray, chronicle, and define their experiences of 'success' in their profession on Facebook groups?

The sub-questions include:

- How do they align their ongoing 'successful' experiences with the Australian Professional Teacher Standards (AITSL, 2018)?
- What supports, such as policies, mentoring, resources, practices, and dispositions, do they cite as enabling and inspiring them to feel successful?
- How does their Facebook group participation contribute towards their identity as out-of-field teachers?
- Does "teacher over-confidence in teaching" (Brown et al., 2022, p. 295), competence, and confidence feature in their reflections about their performance over time, and if so, how?

Positionalities under the spotlight

It is important to be candid about my voice and biases in this project, my researcher positionality. Drawing on Holmes (2020, p. 1), I would describe my positionality as comprising my worldview (ontological assumptions), my beliefs about knowledge (epistemological assumptions), and what underpins my understanding of 'human nature and agency'. Ontologically,

I ascribe to Creswell and Creswell's (2018) view that multiple participant meanings exist. Therefore, I align with constructivism. In this way, I see multiple paths to out-of-field teachers of STEM experiencing success, due to the diversity of teaching contexts and learners they face. This 'experience' will be unique to each out-of-field STEM teacher. Their individual identities will also play a part. Constructivism links to my epistemological stance as an interpretivist, so I aim to engage in sensemaking about participants' views.

Here, I'm turning the lens back onto my subjectivities, which means I am a reflexive researcher (Braun & Clarke, 2006). It also points me towards a qualitative paradigm (Creswell & Creswell, 2018), which is appropriate considering my research will focus on stories, and individual narratives about practices. As a teacher-researcher-journalist, I continuously boundary cross (Akkerman & Bakker, 2011; Hobbs, 2013) between emic (insiderness) and etic (outsiderness) (Beals et al., 2020), so my positionality will be fluid. Subtexts to my positionality include being a white, working-class, heterosexual female in her mid-50s, who is a second-generation migrant to Australia; a bilingual, who has, figuratively speaking, shapeshifted careers multiple times.

Netnography: A prescriptive yet flexible methodology

Netnography (Kozinets, 2020) is a form of ethnography on the internet that has been employed widely to study online communities. This methodology aims to contextualise those conversations and experiences and focus on meaning (Kozinets, 2020). Netnography also offers windows on "naturally occurring behaviors" by approaching "cultural phenomena in their local contexts" (Kozinets et al., 2014, p. 262). Early applications of this methodology were in marketing and advertising.

Netnography is a "relatively new" area for education researchers (Rochette, 2022, p. 215). Researchers have harnessed it for topics, including higher education (Howard, 2016; Eaton & Pasquini, 2020; de Oliveira, 2021), online language teaching (Kulavuz-Onal, 2015; Neuschafer, 2022), primary school homework (Lehner-Mear, 2020), remote learning (Nasu, 2021), early education (Wallace et al., 2018), graduate students (Bumbalough, 2016), doctoral students' loneliness (Janta et al., 2014), and high school mathematics teaching (Ibrahim, 2019). Ibrahim's (2019) foray into a 13,000-teacher-strong Facebook group of mathematics educators is the closest to my research project. She concludes that teachers' participation in the group allowed them to exceed the in-service teacher training standards of the Turkish Ministry of National Education in-service teacher training (Ibrahim, 2019). Her work confirms my assumption that Facebook will be a useful site for my research, as does my involvement in Australian Facebook groups for teachers.

On my netnographic quest, I plan to be a social-media "immersed researcher" (Kozinets & Gambetti, 2021, p. 7), a participant observer on particular teacher Facebook groups. Kozinets continues that the purpose of netnography is to "form cultural understandings about language, power, identity, and desire in the worlds where technology and the social intersect" (2021, p. 7). My approach in harnessing netnography acknowledges Australia's Data-Enabled Research Future: Synthesis Report's (ACOLA, 2022, p. 4) claim that social media "offers unprecedented observations of societal dynamics". In the next section, I will also delve into how using netnography may pose ethical challenges and limitations on my research path, and how these can be addressed.

Netnography offers social media researchers 25 research practices, i.e., instructions, with the flexibility to harness, modify and mix how they collect, analyse and interpret data (Kozinets, 2020, p. 7). Kozinets' most up-to-date instructions split data collection from integration methods, i.e.:

Data collection operations of:

- investigation (simplification, search scouting, selecting, and saving);
- interaction (interview, involvement, and informed consent); and
- immersion (reconnoitering, recording, researching, and reflecting).

Data integration methods of:

- analysis (collating, coding, combining, counting, and charting), and
- interpretation (theming, talenting, totalising, translation, turtling, and troublemaking) (Kozinets, 2020, p. 7).

Often, netnographers curate a collection of online data, expand on it through private reflective journaling, and they interact with online communities there and/or offline. This means that, typically, netnographers employ "investigative, interactive, and immersive procedural operations to produce data" (Kozinets, 2020, p. 193). The researcher *determines* how involved they wish to be. They decide on their stance as an observer or participant observer. The research could take textual posts at face value without engaging with the person who has posted or contact them for follow-up interviews or focus group participation. Or the researcher could do the latter as an active member of the Facebook group, allowing probing to clarify posts and responses (Fontana & Frey, 2005). The researcher may choose to set up a Facebook group specifically to recruit participants, but is aware of the need to "invest the blood, sweat and tears required to nurture emerging communities" (Wallace et al., 2018, p. 1). However, a research-specific closed Facebook group would be more transparent and aligned with ethical best practice because would-be participants would go through an approved consent process before agreeing to join.

An ethical stance

As a researcher choosing to be a participant observer, here is how I plan to ethically observe, interact with, elicit, and collect data from participants on Facebook. I am informed by the ethics guidelines of Callaghan et al. (2021), the Association of Internet Researchers (2019) and the British Psychological Society (2021; 2022) among others. The main premise of my research is that Facebook group posts are *not* in the public domain, so I cannot ethically scrape and extract that data without informed consent from those who post comments on social media.

After securing my university's ethical approval about my process, I will seek consent from administrators of relevant teacher Facebook pages to post an announcement about my research and its inclusion criteria. I have been an active member of several teacher Facebook groups in Australia, such as the Relief Teachers' Association, and, in my home state of NSW, the NSW Teachers' Federation page for casual, relief and temporary teachers, for about a decade. If access is granted, I will monitor posts and comments from then on by self-declared out-of-field teachers of STEM writing about their success. Elements that may indicate this are positive reflections about recent STEM classes, and offering advice to peers, for example. I will privately message out-of-field teachers who have posted on the pages about their success to ask if I can use their comments in my research and clarify if they were (or felt) indeed out-of-field at the time. I would also invite them to a series of Zoom interviews over the following 12 months. During each interview, I will also ask them to draw the elements (on a shared program such as Padlet) that they believe contributed to their ongoing experience of success. Noticing and interpreting narratives of success offers rich opportunities to understand (Quong et al., 1999) perspectives about teaching STEM out- of- field. By using teachers' written posts, the follow-up interview transcripts and the drawings, I expect to have three data sources as a measure of reliability.

If few or no out-of-field teachers agree to participate, I would then post about my success experiences in out-of-field STEM teaching to encourage Facebook group members to comment and share. Plan C would involve setting up my own closed Facebook group, inviting people via other teacher Facebook groups, grey literature sites, LinkedIn, and word of mouth to join. Should those strategies fail, I would refine my scope to either just out-of-field teachers of mathematics or science, or both, approaching relevant Facebook pages for those subject areas. Some of these strategies steer me into auto-ethnography as well as ethnography, possibly emulating the style of Ellis (2022). She harnesses a sociological and communicational lens (Ellis, 2022, p. 4), using literary devices to engage readers in auto-ethnography as a perspective and method (Ellis, 2022, p. 6). This would entail adopting an immersive or "intimate netnography" approach (Patterson & Ashman, 2021, p. 214), which cascades even more levels of ethical sensibilities.

My data collection will comprise interview transcripts and participants' Facebook posts about out-of-field STEM teaching success stories. As a netnographer, I currently vacillate between cloaking or highly cloaking (Kozinets, 2020, p. 184) the data I collect and will need to discuss this with participants. Kozinets (2020) argues that cloaking data takes reasonable safeguards to "conceal the presented identity of social media posters, other research participants, and the precise contents of their traces" (Kozinets, 2020, p. 184). For example, I would name the Facebook groups used for my data collection; paraphrase quotations; change identifiers; and use pseudonyms. Highly cloaking data would go further, making "strong efforts" (Kozinets, 2020, p. 184) to conceal identifiers. For instance, I would *not* name the Facebook group, nor the Australian state I operate in, nor use verbatim quotations, but would alter names and all identifying information. Retaining quotations verbatim in my work easily allows others to reverse search to match those quotations to my research participants, which would breach confidentiality. A higher level of cloaking will demand intense activity in 'cleaning' and de-identifying the data, while retaining sufficient narrative data for my research.

Scoping a route to analysis

Analysis of qualitative data typically follows a spiral pattern (Creswell & Poth, 2018), with multiple iterations to delve deeper into the data, and write and reflect upon it as I collect it. Meanwhile, Stiegletz and colleagues (2018) describe the process of social media analytics as more linear, coursing through "data discovery, collection, preparation and analysis" (p.156). With netnography, researchers can find it challenging working with large volumes of social media data (Stieglitz et al., 2018; Kozinets & Gambetti, 2021). Other issues include the "speed of data creation, variety of data types and questions about data quality" (Stieglitz et al., 2018, p. 158). While these problems may be present in my research, they will not be overwhelming because I will not work with 'big data'. Instead, I will carefully define and maintain my field (Webster & da Silva, 2013, p. 125) to collect data for quality rather than volume. I will initially sweep through the data using netnography processes outlined above. For a more nuanced understanding of where teachers are quoting others (such as school leaders), I will use Kula and Grzelka (2002, pp. 82-83) as a guide for dividing quotation between direct (direct speech, text islets, such as snippets, pseudo-quotations) and indirect (indirect speech, narrated speech) quotations.

Next, I will use the analytical methodology, Reflexive Thematic Analysis (RTA) (Braun & Clarke, 2021). Before selecting RTA, I considered other approaches listed in Stieglitz and colleagues' (2018) Social Media Analytics framework. These alternatives focus on statistical, social network, sentiment, content, or trend analysis (Stieglitz et al., 2018, p. 159). However, as an inductive qualitative analytical method, RTA allows me to work with the data in a theoretically flexible and interpretative way (Braun & Clarke, 2021). I also have recent experience using this methodology. I will use RTA to do an estimated three rounds of hand coding of the textual and diagrammatic to identify patterns of categories that I can coalesce into themes or concepts (Saldana, 2021) and word clouds. In their systematic mapping of netnographic research, Callaghan and colleagues (2021) found that thematic analysis allows the identification of "emergent attitudes and influences" (p.1). This may align with Rochette's (2022) notion of teacher's 'secret stories', that may not feature in public discourse. It is also important for the research to not merely describe and summarise what someone has posted or uttered in interviews, and their frequency, but to develop "nuanced approaches that provide richer insights" (Callaghan et al., 2021, p. 19). A more sophisticated analysis, therefore, involves detecting what has been implied, assumed, as well as missing from the narratives, perhaps even speculating why. Quirke (2022, p. 106) suggests that educators' "pre-existing images" and "taken-for-granted beliefs" are heightened while making sense of and teaching out-of-field. I look at the narratives to illuminate out-of-field STEM teachers' self-understandings.

Success as a malleable construct

Through this analysis, I will refine the conceptual construct of 'success', which poses as problematic at the start of this project. I borrow the parlance 'success story' from my copywriting and content marketing writing work. Akin to a case study, it involves building a coherent narrative, initially setting the scene describing a customer 'pain point' or problem, detailing how they attempted to address that, how they chose the 'winning' solution, ideally my client's service or product. The results are quantified by time and cost savings, efficiency gains, and productivity boosts. This template requires more sophistication when applied to teaching, schools, and out-of-field teachers. A key issue is 'success' for whom, how, why, and where. I will not set parameters to pre-empt what out-of-field STEM teachers signify as their success, but I have a working notion of this. Such teachers might delineate their success by sequence of lessons, consistently engaging students in learning and/or quality learning outcomes, even feedback from others in the school community, for example.

Out-of-field teachers may measure success against what they perceive success to look and feel like, informed by their in-field discipline area, argues Hobbs (2013). Therefore, success may be a very subject-specific experience. In collecting my data, I am mindful that I may differentiate 'success' further, such as between ordinary successes, i.e., 'small wins', and what Quinn et al. (2021, p. 15) denote as "exceptional success". Such high-level success is "unusually positive performance that exceeds typical performance" and goes beyond "what previously seemed possible" (Quinn et al., 2021, p. 17). I would regard an out-of-field teacher perceiving they have achieved exceptional success as presumably having consolidated their professional development and experience to demonstrate competence and confidence. In this way, I create a distinction between a 'lesson that went well' and success, which is more enduring. I assume, as well, that out-of-field teachers will acknowledge that they are out-of-field and be comfortable about reflecting on their practice, which may not be the case (Hobbs, 2013).

The data collected will help define and elaborate upon teachers' perspectives/conceptualisations of success possibly through the lens of capabilities, pathways and identities (Hobbs et al., 2022), signature pedagogies (Shulman, 2005), teaching 'near field', across subject boundaries (Hobbs & Porsch, 2022), and temporality to test Du Plessis' (2020) anecdotal finding that out-of-field teachers felt competent within three-to-five years. My research will also adopt Hobbs' (2020) categories that out-of-field educators used to teach science, i.e. "content, teaching strategies, students and their learning, the school landscape and the professional self" (p. 725), and indeed capturing teachers' pedagogical and personal response to out-of-field work (Hobbs, 2020, p. 741). From Du Plessis et al. (2019, p. 223), I intimate possible criteria for success in teachers' enjoyment and confidence in teaching out-of-field. Rochette (2022, p. 112) states that out-of-field teachers can "re-position themselves as increasingly agentic professionals" when they can use their professional judgement and have support when needed. Hence, a way forward could be to consider 'success' a 'working' term or malleable construct to evolve as I continue to engage with scholarly literature and eventually my data sample.

Hence, my research 'travels' in this project will follow this methodological map, which unfurls from sociocultural theories and models of identity, and the theory of practice architectures. The methodology, that is route, will be innovative because few scholars have taken the netnography path to explore a subset of Australian teacher Facebook groups to focus on members posting about teaching STEM successfully out-of-field over time and, from the latter's perspective. My aim is to foreground such narratives of success, drawing themes from them for interpretation and analysis, and this constitutes a strengths-based approach to further understand the out-of-field teaching phenomenon. My research may therefore offer some nuance, perhaps not so much balance, to the predominantly deficit discourse in the public sphere and scholarly work about this phenomenon. Importantly, my findings will be contextualised, so will not purport to be generalisable.

Acknowledgements

The author is grateful for the guidance of her main supervisor, Associate Professor Linda Hobbs, secondary supervisor, DECRA Scholar Dr Emma Rowe, both from Deakin University, and supplementary supervisor, Associate Professor Peter Simmons, from Western Sydney University.

References

ACOLA. (2022). Australia's data-enabled research future: Synthesis report. www.acola.org

Akkerman, S. F., & Baker, A. (2011). Boundary crossing and boundary objects. *Review of Education Research, 81*(2), 132-169. https://doi.org/10.3102/0034654311 404435

Association of Internet Researchers. (2019). Internet research: Ethical guidelines 3.0 DRAFT.

Audit Office of New South Wales. (2019). *Supply of secondary teachers in STEM-related disciplines.* NSW Auditor-General's Report, Performance Audit, 29 January.

Australian Curriculum. (n.d.). STEM in the Australian Curriculum. Australian Curriculum and Reporting Authority.

Australian Curriculum Reporting Authority (ACARA). (2020). *Guide to understanding the Index of Community Socio-educational Advantage (ICSEA): My School.*

Australian Government Department of Education. (2023). National STEM school education strategy 2016-2026.

Australian Institute of Teaching and School Leadership Limited (AITSL). (n.d.). *Australian teacher workforce data.*

Australian Institute of Teaching and School Leadership Limited. (2018). *Australian Professional Standards for Teachers*. Australian Institute for Teaching and School Leadership.

Australian Institute of Teaching and School Leadership Limited. (2022). *Australian teacher performance and development framework.* Australian Institute for Teaching and School Leadership.

Baker, J. (7 October 2021). Public school teacher shortage raises fears they will 'run out of teachers'. *The Sydney Morning Herald.*

Baker, J. (2 May 2022). Teachers warn plan to address staff shortages could lead to lower standards. *The Sydney Morning Herald.*

Baroutsis, A. (2019). Mapping the field of education research and the media. In A. Baroutsis, S. Riddle, & P. Thomson (Eds.), *Education research and the media: Challenges and possibilities*. Routledge.

Beals, F. M., Kidman, J., & Funaki, H. (2020). Insider and outsider research: Negotiating self at the edge of the emic/etic divide. *Qualitative Inquiry, 26*(4), 593-601. https://doi.org/10.1177/1077800419843950

Beijaard, D., Meijer, P. C., & Verloop, N. (2004). Reconsidering research on teachers' professional identity. *Teaching and Teacher Education*, 20, 107-128.

Beswick, K., Fraser, S., & Crowley, S. (2016). "No wonder out-of-field teachers struggle!" Unpacking the thinking of expert teachers, *Australian Mathematics Teacher, 72*(4), 16-20.

Bourdieu, P. (1977). *Outline of a theory of practice.* Cambridge University Press.

Braun, V., & Clarke, V. (2006). Using thematic analysis in psychology. *Qualitative Research in Psychology. 3*(2), 77-101. https://doi.org/10.1191/1478088706qp063oa

Braun, V., & Clarke, V. (2021). *Thematic analysis: A practical guide.* SAGE.

British Psychological Society. (2021). *Ethics guidelines for internet-mediated research.*

British Psychological Society. (2022). Supplementary guidance on the use of social media. www.bps.org.au

Brown, J. P., Sawatzki, C., & Widjaja, W. (2022). Tensions between theory and practice: A mathematics education perspective. In P. J. White, R. Tytler, J. P. Ferguson, & J. Cripps Clark (Eds.), *Contemporary approaches to research in STEM education: Methodological approaches to STEM education volume 3* (pp. 286-307). Cambridge Scholars Publishing.

Bumbalough, M. (2016). The impact of social media on identity formation: A netnographic study of Korean graduate students' virtual communities of practice (doctoral dissertation). Indiana University. ProQuest Dissertations Publishing.

Caldis, S. (2017). Teachers having to know what they do not know. *Geography Bulletin*, *49*(1),13-16.

Callaghan, K., Mehmet, M., Simmons, P., & Curley, B. (2021). Policy insight through social and online media: A systematic mapping of qualitative passive Citizensourcing Studies. *The Journal of Social Media in Society, 10*(2), 1-26.

Clandinin, D. J., & Connelly, F. M. (1996). Teachers' professional knowledge landscapes: Teacher stories, stories of teachers, school stories, stories of schools. *Educational Researcher*, *25*(3), 24-30.

Conant, J. (1963). *The education of American teachers.* McGraw-Hill, in R. M. Ingersoll (1999), The Problem of Underqualified Teachers in American Secondary Schools. *Education Researcher, 28*(2), 26-37.

Creswell, J. W., & Creswell, J. D. (2018). *Research design: Qualitative, quantitative & mixed methods approaches* (5th edition). SAGE.

Creswell, J. W., & Poth, C. N. (2018). *Qualitative inquiry and research design: Choosing among five approaches* (4th edition). SAGE.

Darragh, L. (2016). Identity research in mathematics education. *Educational Studies in Mathematics, 93*(1),19-33.

de Oliveira, E. T. (2021). Assessment, interaction and technology in distance education: A netnographic study at a Brazilian virtual university. *Open Praxis*, *13*(3), 242-252. http://doi.org/10.5944/openpraxis.13.3.133

DFAT (Department of Foreign Affairs and Trade). (2019). *Education sector workforce planning: Practitioner level.* Education Learning and Development Module, Diplomatic Academy, DFAT Australian Government.

Du Plessis, A. E. (2020). *Out-of-field teaching and education policy: international micro-education policy*. Springer.

Du Plessis, A. E., Hobbs, L., Luft, J. A., & Vale, C. (2019). The out-of-field teacher in context: The impact of the school context and environment. In L. Hobbs, & G. Törner (Eds.), *Examining the phenomenon of "teaching out-of-field"* (pp. 217-242). Springer.

Du Plessis, A. E., & McDonagh, K. (2021). The out-of-field phenomenon and leadership for wellbeing: Understanding concerns for teachers, students and education partnerships. *International Journal of Education Research*, *106*, 101724. https://doi.org/10.1016/j.ij34.2020.101724

Eaton, P. W., & Pasquini, L. A. (2020). Networked practices in higher education: A netnography of the #AcAdv chat community. *The Internet and Higher Education*, *45*, 100723. https://doi.org/10.1016/j.iheduc.2019.100723

Ellis, C. (2022). *Revision: Autoethnographic reflections and life and work.* Routledge.

Fontana, A., & Frey, J. H. (2005). The interview. In N. K. Denzin, & Y. S. Lincoln (Eds.), *The Sage handbook of qualitative research* (3rd ed., pp. 695-727). Sage.

Finkel, A. (2020). *The Finkel files: Selected speeches from Australia's Chief Scientist Dr Alan Finkel AO 2016-2020*. Office of the Chief Scientist.

Hatano, G., & Inagaki, K. (1986). Two courses of expertise. In H. Stevenson, H. Azuma, & K. Hakuta (Eds.), *Children development and education in Japan* (pp. 262-272). Freeman.

He, Y. (2019). Strengths-based mentoring in pre-service teacher education: A literature review. *Mentoring & Tutoring: Partnership in Learning. 17*(3), 263-275.

Hobbs, L. (2013). Teaching 'out-of-field' as a boundary-crossing event: Factors shaping teacher identity. *International Journal of Science and Mathematics Education. 11*(2), 271-297. https://doi.org/10.1007/s10763-012-9333-4

Hobbs, L., Campbell, C., Delaney, S., Speldewinde, C. and Lai, J. (2020). *Defining and mapping out-of-field teaching in Victorian government schools*, Deakin University, Victoria, Prepared for the Victorian Department of Education and Training.

Hobbs, L, Du Plessis, A. E., Oates, G., Caldis, S., McKnight, L., Vale, C., O'Connor, M., Rochette, E., Watt, H., Weldon, P., Richardson, P. & Bateup, C. (2022). *Australian national summit on teaching out-of-field: Synthesis and recommendations for policy, practice and research.* Deakin University and Griffith University.

Hobbs, L., Du Plessis, A. E., Quinn, F., & Rochette, E. (2019). Examining the complexity of the out-of-field teacher experience through multiple theoretical lenses. In L. Hobbs, & G. Törner (Eds.), *Examining the phenomenon of "teaching out-of-field': International perspectives of teaching as a non-specialist* (pp. 87-128). Springer.

Hobbs, L., & Porsch, R. (2022). *Out-of-Field Teaching Across Teaching Disciplines and Contexts.* Springer.

Hobbs. L., & Törner, G. (Eds.) (2019). *Examining the phenomenon of "Teaching Out-of-field": International perspectives of teaching as a non-specialist*. Springer.

Holmes, A. G. D. (2020). Researcher positionality – A consideration of its influence and place in qualitative research – A new researcher guide. *Shanlax International Journal of Education*, *8*(4), 1-10, https://doi.org/10.34293/education.v8i4.3232

Howard, L. (2016). Casting the 'net' in autonetnography: Professional development in networked learning teaching praxis. *Proceedings of the European Conference on e-Learning, Learning, Education and Development*, Cabinet Office.

Ibrahim, Y. (2019). Using Facebook groups to support teachers' professional development. *Technology, Pedagogy and Education*, *28*(5), 589-609. https://doi.org/10.1080/1475939X.2019.1686714

Ingersoll, R. M. (1999). The problem of underqualified teachers in American secondary schools. *Educational Researcher*, *28*(2), 26-37. http://www.jstor.org/stable/1177187

Jakovac, M. (2022). Hey, Miss. Are you a real teacher? Australian media representations of teaching STEM out-of-field (unpublished dissertation). Deakin University, Australia.

Janta, H., Lugosi, P., & Brown, L. (2014). Coping with loneliness: A netnographic study of doctoral students. *Journal of Further and Higher Education*, *38*(4), 553-571. https://doi.org/10.1080/0309877X.2012.726972

Kayan-Fadlelmula, F., Sellami, A., Abdelkader, N., & Umer, S. (2022). A systematic review of STEM education research in the GCC countries: Trends, gaps and barriers. *International Journal of STEM Education*, *9*(2), 1-24. https://doi.org/10.1186/s40594-021-00319-7

Kemmis, S. (2022). *Transforming practices: Changing the world with the theory of practice architectures*. Springer.

Kendall, N., & Thangaraj, M. (2013). Ethnography. In A. A. Trainor, & E. Graue, E. (Eds.), *Reviewing qualitative research in the social* sciences (pp. 87-107). Routledge.

Kozinets, R. V. (2020). *Netnography: The essential guide to qualitative social media research*. SAGE.

Kozinets, R. V., Dolbec, P., & Earley, A. (2014). *Netnographic analysis: Understanding culture through social media data*. SAGE. https://dx.doi.org/10.4135/9781446282243

Kozinets, R. V., & Gambetti, R. (Eds.) (2021). *Netnography unlimited: Understanding technoculture using qualitative social media research.* Routledge.

Kula, A. M., & Grzelka, M. (2022). Quotation in social media: How sharing other people's words could increase misinformation. *Studia Humanistyczne AGH*, *21*(2), 81-98. https://doi.org/10.7494/human.2022.21.2.81

Kulavuz-Onal, D. (2015). *Using netnography to explore the culture of online language teaching communities*. Computer Assisted Language Instruction Consortium, CALIDCO Journal (Online) ISSN 2056-9017. https://doi.org/10.1558/cj.v32i3.26636

Kurup, P. M., Yang, Y., Li, X., & Dong, Y. (2021) Interdisciplinary and integrated STEM. *Encyclopedia* 1, 1192-1199. https://doi.org/10.3390/encyclopedia1040090

Laming, A. (2022). [Committee Chair]. Standing Committee on Employment, Education and Training Inquiry into innovation and creativity: Workforce for the new economy Innovation and creativity, interview with author, 20 April, 2022.

Lehner-Mear, R. (2020) Negotiating the ethics of Netnography: Developing an ethical approach to an online study of mother perspectives. *International Journal of Social Research Methodology*, *23*(2),123-137. https://doi.org/10.1080/13645579.2019.1634879

Lund, K., Jeong, H., Grauwin, S. & Jensen, P. (2020). Research in education draws widely from the social sciences and humanities. *Frontiers in Education*. *5*(2020), 1-20. https://doi.org/10.3389/feduc.2020.544194

Milner-Bolotin, M. (2018). Evidence-based research in STEM teacher education: From theory to practice. *Frontiers in Education*. *3*(92), 1-9. https://doi.org/10.3389/feduc.2018.0092

Mockler, N. (2022). What we really mean when we talk about teacher quality. *Australian Association for Research in Education (AARE),* May 23.

Mohamad Hasim, S., Rosli, R., Halim, L., Capraro, M. M., & Capraro, R. M. (2022). STEM professional development activities and their impact on teacher knowledge and instructional practices. *Mathematics*, *10*(7), 1109.

Nadelson, L. S., Callahan, J., Pyke, P., Hay, A., Dance, M., & Pfester, J. (2013). Teacher STEM perception and preparation: Inquiry-based STEM professional development for elementary teachers. *The Journal of Educational Research, 106*(2), 157-168. https://doi.org/10.1080/00220671.2012. 667014

Nasu, V. H. (2021). Remote learning under the COVID-19 social distancing: Reflections and a netnographic study. *Journal of Information Systems and Technology Management, 18*. https://doi.org/10.4301/S1807-1775202118005

NESA (NSW Education Standards Authority). (2021). *Subject content knowledge requirements abridged*. NSW Education Standards Authority, NSW Government, 1-21.

Neuschafer, T. (2022). A netnographic analysis of how Duolingo supports German and Spanish learning needs. *Digital Culture and Education, 4*(3),1-16.

O'Connor, M., & Thomas, J. (2019). *Australian secondary mathematics teacher shortfalls: A deepening crisis*. Australian Mathematical Science Institute, AMSI Occasional Paper 2, May.

Pacana, N. M. S., Ramos, C. D., Catarata, M. N., & Inocian, R. B. (2019). Out-of-field social studies teaching through sustainable culture-based pedagogy: A Filipino perspective. *International Journal of Education and Practice*. 7(3), 230-241. https://doi.org/10.18488/journal.61.2019.73.230.241

Parliament of Australia. (2017). *Innovation and creativity: Inquiry into innovation and creativity: Workforce for the new economy, final report into the House of Representatives Standing Committee on Employment, Education and Training*. Commonwealth of Australia, May.

Patterson, A. & Ashman, R. (2021). Getting Up-Close and Personal with Influences: The Promises and Pitfalls of Intimate Netnography. In R. V. Kozinets & R. Gambetti. (Eds.), *Netnography Unlimited: Understanding Technoculture Using Qualitative Social Media Research*. pp. 217-240. Routledge.

Porsch, R. & Whannell, R. (2019). Out-of-field teaching affecting students and learning: What is known and unknown (pp. 179-191). In L. Hobbs & G. Törner (Eds.), *Examining the phenomenon of 'teaching out-of-field': International perspectives of teaching as a non-specialist*. Springer.

Quin, R. W., Myers, C. G., Kopeman, S., & Simmons, P. (2021). How did you do that? Exploring the motivation to learn from others' exceptional success. *Academy of Management Discoveries*. 7(1), 15-39. https://doi.org./10.546/amd.2018.0217

Quirke, S. (2022). Teaching mathematics out-of-field and the journey to obtaining mathematics teacher certification. In E. Costello, P. Girme, D. Hyland, T. Kaur, O. Kelly, T. McLoughlin, & P. Van Kampen (Eds.), *Proceedings of the CASTeL 9th STEM Education Research Conference* (pp. 100-109). Dublin City University. https://doi.org/10.5281/zenodo.6778381

Quong, T., Walker, A., & Bodycott, P. (1999). Exploring and interpreting leadership stories. *School Leadership & Management, 19*(4), 441-53. November.

Rochette, E. (2019). Teaching geoscience out-of-field with digital technologies: Understanding agency through positioning theory (unpublished dissertation). University of Melbourne, Australia. https://doi.org/10.1/11343/241644

Rochette, E. (2022). An ethical dilemma inspiring exploration into netnography. In P. J. White, R. Tytler, J. P. Ferguson, & J. Cripps Clark (Eds.), *Contemporary approaches to research in STEM education: Methodological approaches to STEM education Volume 3*. Cambridge Scholars Publishing.

Saldaña, J. (2021). *The coding manual for qualitative researchers*. SAGE.

Shernof, D. J., Sinha, S., Bressler, D. M., & Ginsburg, L. (2017). Assessing teacher education and professional development needs for the implementation of integrated approaches to STEM education. *International Journal of STEM Education*, *4*(1), 1-16.

Shulman, L. (1986). Those who understand: Knowledge growth in teaching. *Educational Research*, *15*(2), 4-14. doi.org/10.2307/1175860

Shulman, L. (2005). Signature pedagogies in the professions. *On Professions & Professionals, 235*(3), 52-59. https://www.jstor.org/stable/20027998

Siegel, D., & Giamellaro, M. (2022). Non-STEM teachers finding their place in STEM. *Journal of Science Teacher Education*, *33*(6), 579-597. https://doi.org/10.1080/1046560X.2021.1968992

Spaargaren, G., Lamers, M., & Weenink, D. (2016). Introduction: Using practice theory to research social life. In *Practice Theory and Research.* Routledge. https://doi.org/10.4324/978131565690-11

Stieglitz, S., Mirbabaie, M., Ross, B., & Neuberger, C. (2018). Social media analytics–Challenges in topic discovery, data collection and data preparation. *International Journal of Information Management. 39*(218), 156-168. https://doi.org/10.1016/j.ijinfomgt.2017.12.002

Thomson, P., & Riddle, S. (2018). Who speaks for teachers? Social media and teacher voice. In *Education Research and the Media* (pp. 119-134). Routledge.

Trainor, A.A. & Graue, E. (Eds.), *Reviewing qualitative research in the social* sciences (pp. 87-107). Routledge.
Vale, C. (11 February 2019). How school leaders can support out-of-field teachers. *Teachspace blog,* Monash University.
Vale, C., & Drake, P. (2019). Attending to out-of-field teaching: Implications of and for education policy. In L. Hobbs & G. Törner (Eds.), *Examining the phenomenon of 'teaching out-of-field': International perspectives of teaching as a non-specialist*. Springer.
Vale, C., Hobbs, L., & Speldewinde, C. (2022). Challenging the representations and assumptions of out-of-field teaching. In L. Hobbs & R. Porsch (Eds.), *Out-of-field teaching across disciplines and context*. Springer.
Wallace, R., Costello, L., & Devine, A. (2018). Netnographic slog: Creative elicitation strategies to encourage participation in an online community of practice for early education and care. *International Journal of Qualitative Methods*, *17*(1), 1-13. https://doi.org/10.1177/1609406918797796
Webster, J. P., & Marques da Silva, S. (2013). Doing educational ethnography in an online world: Methodological challenges, choices and innovations. *Ethnography and Education*, *8*(2), 123-130. https://doi.org/10.1080/17457823.2013.792508
Weldon, P. R. (2016). Policy insights: Out-of-field teaching in Australian secondary schools. Australian Council for Educational Research.
Weldon, P. (2018). Early career teacher attrition in Australia: Evidence, definition, classification and measurement. *Australian Journal of Education*. *62*(1), 61-78. https://doi.org/10.1177/0004944117752
Wenger, E. (1998). *Communities of practice: Learning, meaning, and identity*. Cambridge University Press.
Williams, P. J. (2019, March). The principles of teaching and learning in STEM education. In *AIP Conference Proceedings* (Vol. 2081, No. 1, p. 020001). AIP Publishing LLC.

Margaret Jakovac (BArts (Journalism), Grad Dipl Comm Mgmt, MTeach (Primary), Grad Cert Ed Research) - Deakin University, School of Education, 75 Pigdons Rd, Waurn Ponds, 3216, Victoria, Australia.

Margaret Jakovac is a distance and part-time PhD student exploring netnography for success stories of out-of-field STEM high-school teachers. Under the surnames Paton and Ryan, she has written extensively about K-12 teaching practices for Australian education publications. Since 2011, Margaret has done occasional stints as an out-of-field STEM teacher. She also hosts the Out-of-Field Teaching Toolkit podcast. Margaret was jointly

awarded a Walkley Foundation grant in 2019 to research and write about the implications of school fundraising. She was *The Sunday Age*'s first journalism cadet.
https://www.linkedin.com/in/margaretpatonwriterbusinessenglishcoach/

CHAPTER 10

EXPERIENTIAL KNOWING WITH THE MORE-THAN-HUMAN WORLD

SANDRA WOOLTORTON & PETER REASON

Abstract

Co-operative inquiry emerged in the 1970s, recognising the impossibility of conducting authentic research in the social sciences without engaging with persons as aware, self-directing beings. The approach asserts that people have a right to contribute to knowledge that purports to be about them. It acknowledges that because it is politically empowering, this can lead to political struggle. Lately, it has become clear that these arguments can and should be extended to apply to the more-than-human world, thus re-centering previously marginalised relational knowledges and voices. Recent work has focused particularly on the agency of place, such as River. In this chapter, we reflect upon the philosophy and practice of co-operative inquiry for life, in re-engagement with the more-than-human world. Most modern Western humans have been socialised to draw a distinct line between humans and other-than-human, to see a world open to exploitation. In contrast, we draw on living cosmos panpsychism (Mathews, 2003, 2019) to see a world in which all beings, including humans, are part of an intelligent and evolving cosmos. We seek experiential, conceptual, creative, and practical responses to questions such as: *What would it be like to live in a world of sentient beings rather than inert objects?*

Key words: Co-operative inquiry, More-than-human world, Agency of place, Living cosmos panpsychism

Opening

> Zen masters teach us not to seek the extraordinary, not to look for special or 'sacred' places. To see that which is special stops us from seeing what is before our eyes: the specialness of the everyday, how everything rolls together in being and non-being, how we are every moment part of a living planet. (Reason, 2017, p. 203)

The point of this chapter is to learn how to pay attention; to notice relationships and the agency[1] of the other-than-human world. Through the intimate human association nurtured by the practice of co-operative inquiry, we can learn to look and look again in association with others, allowing an occluded view of the world to open. This is a living, sentient[2] world, in Thomas Berry's words, a community of subjects, not a collection of objects (1988). This world is not only sentient but also communicative; a community in mutual contact and communication, co-creating a 'poetic ecology'. We, humans, bring to it–or can bring to it, if we choose–an attention that calls it forth on a new expressive plane, a plane of meaning and not merely of causation.

We need to re-learn how to do this. Many moderns live in a world alienated from relationships and relational knowing; they see the world as separate objects, as brutish matter (Ghosh, 2021). This worldview disavows the body and the fundamental interconnectedness of knowing (Abram, 1996; Mueller, 2017). It is arguably at the root of the ecological catastrophe that is currently unfolding, now faster than predicted. This is evidenced by the biodiversity crises across the planet (Díaz et al., 2019); the 2022 devastating floods in Pakistan described by the United Nations as climate carnage (United Nations, 2022); and many more unnatural disasters. As a case in point, the state of the Australian environment is poor and the outlook is deteriorating (Cresswell et al., 2021). The Australian climate has warmed by about 1.5°C, sea surface temperatures have increased, extreme fire weather has increased, and river and stream flows have decreased in most

---

[1] For us, agency means to oversee one's own life, and to be capable of influencing others. This applies to the more-than-human world as well, in the sense of being able to respond, affect or independently impact place or others.

[2] By sentient or sentience, we refer to a capacity to feel, experience and/or consciously act. We recognise that individuals, places, and the cosmos are sentient, and that perhaps our cultural conditioning has prevented this awareness. "Eventually you may feel the entire planet nurturing you with life energy." These are the words of Harding, S. (2022). *Gaia alchemy: The reuniting of science, psyche and soul.* Inner Traditions Bear and Company. Sentience relates to sensing and sensation.

regions across the continent since 1975 (CSIRO, 2022). Some say these calamities are consequences of humans learning to think in ways that privilege theory and abstract knowledge, even while ignoring the consequences of these actions and one's own experiences, observations, feelings, values, and sentiments (Mathews, 2017b).

We briefly explain. In post-Cartesian philosophy there is a schism between rationality and nature. This is part of a mind/matter dualism, a metaphysic that regards both the human body and 'nature' as exterior (and inferior) to an internalised mind (Plumwood, 1993). Using this logic, the mind comprehends one's body and world from a separate, internal location. For this reason, Westerners may not recognise contrasting evidence–a bird visitation in response to a call to River, perhaps–because the world around us is seen as non-responsive; and bodily experiences, feelings, and senses are trivialised or relegated. It seems to us that *un*learning is required, along with removal of the modern narrative of human exceptionalism.

In this chapter, we seek experiential, conceptual, creative, and practical responses to questions of the kind: What would it be like to live in a world of sentient beings rather than inert objects? How would we relate to such a world? How might we listen to rivers, and learn to relate in ways that reveal ancient knowledges often occluded by the mechanisms of the West? What are the implications of these questions, observations, and responses for education?

We briefly review the environmental humanities to explore practices of noticing, before investigating Freya Mathews' (2021) living cosmos panpsychism in which all beings, including humans, are part of an intelligent and evolving cosmos. We introduce the practice of co-operative inquiry, an approach to living life as an inquiry for deepening engagement within local more-than-human[3] worlds. We offer journal entries of our daily practice, to illustrate the wonder in and of the everyday, after which we draw methodological and educational implications.

Noticing

Writers in the environmental humanities cultivate the arts of noticing–to observe, attend to, listen, smell, sense, and otherwise fully engage (Harding, 2009; Tsing, 2021). To closely attend is to recognise the world anew, to

[3] In this and other contexts, we use the term more-than-human to avoid notions of human exceptionalism. It is to include humans within a community of beings.

recultivate a sense of wonder–or perhaps a sense of horror. It is to bring entanglements to the experiential, visual/imaginal, and practical realms of existence, rather than just the theoretical (Van Dooren, 2022). It is to be fully human in a more-than-human world (Mueller, 2017). Marshall (2001) writes about paying attention, but not with an ecological focus. She writes, "I work with a multi-dimensional frame of knowing; acknowledging and connecting between intellectual, emotional, practical, intuitive, sensory, imaginal and more knowings" (2001, p. 434). Below, we offer a narrative of noticing, to bring to focus different senses and ways of becoming reacquainted with places we live with.

Tsing (2021) notices networks of all kinds including patterns of unintentional coordination apparent in assemblages. She thinks-with matsutake mushrooms, the most valuable in the world, being a weed that grows in disturbed forests and environmental messes in the Northern hemisphere. For her, matsutake suggests attention to seasons and precarious living in the present. In describing a visit to the tanoak forests of coastal Oregon, where stumps of logged Douglas fir provide likely matsutake habitat, tanoak leaves covered the ground like a rug, and she commented on the seeming impossibility of finding mushrooms underneath. She was shown how to feel the leaves with her hands alone, "a new way to learn the forest" (Tsing, 2021, p. 278), illustrating the depths of awareness which many of us may acquire.

Mueller's (2017) observations and description of the deep interrelationship between humans and salmon, causes him to wonder if working to save the salmon might be the salmon working to save the human; and whether earth might be experiencing itself through so many of these different kinds of embodied awareness, be they dragonfly, Sitka spruce, salmon, human, or whale. To recognise interrelationships not previously noticed, may enable creative ways of extending the act of knowing.

Van Dooren (2022) notes that in this era of colonisation and extinctions, snails rarely get a mention in the West, even though they are intricately interwoven with Hawaii tradition and cultural practice. He writes snail stories to animate a sense of wonder for snails and their worlds, to acknowledge lives at the edge of extinction; and to face the unravelling and deep wounding taking place at this point in history. While he was visiting Hawaii one time in 2019, the last snail (named George) of the species *Achatinella apexfulva*, passed away. For him, the point of paying close attention is to learn to observe, notice relationships, respond, inspire to transform, and perhaps understand the world afresh.

A five-day meeting of artists, musicians, environmentalists, and Indigenous custodians took place at a remote cattle station in central Australia, a site Mathews (2009) refers to as utterly beautiful. Referring to the event, she writes about Dreaming stories for the specific Country[4] being told by Indigenous custodians, and an ethnobotanist speaking lovingly of each plant he pointed out while leading a walking tour. She says at several points, different individuals noticed their activities forming poetic scenarios or narratives they had not previously imagined, yet these were perfectly suited to the circumstances. In other words, place-love and respect can bring a communicative response. For example:

> Each time we came to a gap in the proceedings, it was as if the land, or the world, stepped in, and offered poetic comments or denouements that exceeded anything we could have devised. There was a breathtaking display of lightning, for instance, just at a particular 'flat spot' in the women's final performance of place; multitudes of actual frogs gathered on the slope leading down to the riverbed just after the men had performed a frog dance; crucial conversations were punctuated, at just the right moments, with expressive bird calls; … It felt like a coming alive of the world, a flow of configurations of circumstances along axes of meaning. (Mathews, 2009, p. 2)

This narrative illustrates the significance of beginning with open-minded sensory observations–of what is in front of us. We need to recognise those things we learned *not* to notice because those experiences may not fit the Western 'everyday' worldview or its explanations. We need to learn to feel, see or hear previously unimagined possibilities, especially those we may have blocked out because of theories or beliefs we already hold. This calls for a philosophical exploration to consider the nature of reality.

Living cosmos panpsychism

As guidance for a series of co-operative inquiry ventures, we have drawn on Freya Mathews' 'living cosmos panpsychism' (2021) through which we can understand all beings, including humans, as part of a sentient, evolving cosmos. An important dimension of this approach is that instead of attempting to understand the world and *representing* it with theory, panpsychism shows that our task is to *participate with* the world to make meaning. Mathews teaches that the challenge for a philosopher is not how

---

[4] In Australia, when capitalised Country refers to a more complex notion than that of the standard usage. It is an Indigenous meaning that includes spirituality, social systems, well-being, kinship, and obligations to care for Country.

to think *about* the world, but to offer perspectives that enable us to *live in* and *participate with* the world (Mathews, 2017c).

In the living cosmos panpsychic perspective, some kind of innerness is as fundamental to the nature of reality as is matter (Mathews, 2003, 2019, 2023). In this sense, there can be no brute matter of the type suggested by physics. We might think of the cosmos as a consistent field of mind-matter, a self-actualising meaning-making whole with a concern for its own continuation and self-increase. It has its own *conatus,* the will or impulse to realise and increase its own existence and reach towards its full potential. (Mathews, 2021). This One field of mind-matter self-differentiates into Many: distinct, self-realising, mutually responsive beings, that is, the gum tree, salmon, snail, mycelium, and you and me, majestic yet impermanent centres of action. Each is a system, a self, as is the cosmos (Mathews, 2023).

Using living cosmos panpsychism, we recognise the cosmos as a communicative, responsive order of meaning/s, unfolding together within the causal, material order. As a community of subjects, the Many reaches for each other in reciprocal communication, co-creating a 'poetic ecology': the foundational erotics of touching the world and being touched by it in return (Mathews, 2021). Participation in this living world of relationality, is a foundational experience of reality. The expression of meaning does not radiate solely from the human side. The world is capable of, actively seeks, engagement with humans, enabling the possibility of a "communicative encounter, of reciprocal presence, a presence that *answers back* when our questions send out tentacles of attention in search of it" (Mathews, 2017a, p. 223). This way, the world inscribes our lives with meanings of its own, offering defining conduits for our agency. Hence all things, including the Earth, are vital to the composition of the *living cosmos*, all the same sentient cloth.

Mathews has developed practical theories of a communicative order in living systems. She names the living, communicative order of things ontopoetics, while the causal order of things is physics. Ontopoetics is, "the study of the poetic order, the poetic meanings that structure the core of things" (Mathews, 2009, p. 15). She contends that the communicative order will structure further interchanges when we choose to participate with those meanings. As well, living cosmos panpsychism has a moral dimension that is relevant to everyday practice and contemporary society. This suggests that we should seek to align ourselves with the performative intelligence in the way things are, and act in accordance with it. In other words, a moral 'Ought' exists at the core of the 'Is' (Mathews, 2021), that is, we 'ought' to act in accord with the communicative order of things. This is a "Law of the living Cosmos", which "permeates the very fabric of existence" (Mathews, 2023, p. 39), a normative imperative. Mathews (2021, 2023) explains that the 'Law of the living cosmos' is particularly discernible in the unfolding of living systems, as a synergistic pattern characterised by the twin principles of conativity or striving for self-actualisation, and mutual accommodation/least resistance. This has great similarities with the Indigenous principle of Law of the Land, with its relational ethos (Mathews, 2023; Poelina et al., 2020).

Mathews asserts that scientific materialism has brought devastating technologies and commodities, which have *un*taught us moderns how to live (2021). We now need to learn how to live anew, in accordance with the Law of the Living Cosmos and the Law of the Land. This requires an agential, reciprocal, direct way of relating-with the world; entirely different to the world mainly known through theoretical abstractions separate from the world. In summary, panpsychism is conducive to an experiential, direct engagement with the world, such as co-operative inquiry, one that is reflective in creative and conceptual ways.

Co-operative inquiry

A co-operative inquiry has two or more persons exploring a topic through their own experience of it, utilising a series of cycles in which they move between their experience and reflecting on it together. The research method began in earnest in the 1970s, after recognising the importance of conducting authentic research in the social sciences while engaging with persons as aware, self-directing beings (Reason, 2002). Co-operative inquiry was originally associated with the humanistic quest for personal and social transformation aligned with the values of autonomy, co-operation and holistic body-mind. It envisioned persons inquiring together in reciprocal relation, using the full range of their sensibilities, into any aspect of the human condition (Reason & Heron, 1997).

Lately, it has become clear that these arguments can and should be extended to apply to the more-than-human world, thus re-centering previously marginalised relational knowledges and voices. Recent work has focused particularly on the agency of place, such as River (Kurio & Reason, 2021; Riley & Reason, in press). There are now many approaches to researching the place-human experiential relationship, and one of us has referred to co-operative inquiry as a post-qualitative research method. As we understand it, this includes and is greater than qualitative research in that it is no longer confined to responses by humans alone (Poelina et al., 2022).

Co-operative inquiry is an iterative process in which co-researchers engage in cycles of action and reflection through an extended epistemology—that is, extended from the rational-empirical categories of traditional research. This epistemology embraces experiential knowing through meeting and encounter; presentational knowing through aesthetic, creative, and expressive forms; propositional knowing through words, concepts, and critique; and finally, participative, or practical knowing in the exercise of diverse skills—which may be physical, interpersonal, attentional, and/or political. Practical knowing is the epitome of prior forms of knowing, involving the whole body-mind. These forms of knowing are brought to bear upon each other through inquiry cycles to enhance their mutual congruence, both within each inquirer and the inquiry group. See Figure 10.1 for a visual explanation of one cycle of fourfold knowing. The process is cycled perhaps many times. We suggest the process is continuous for lifelong learning.

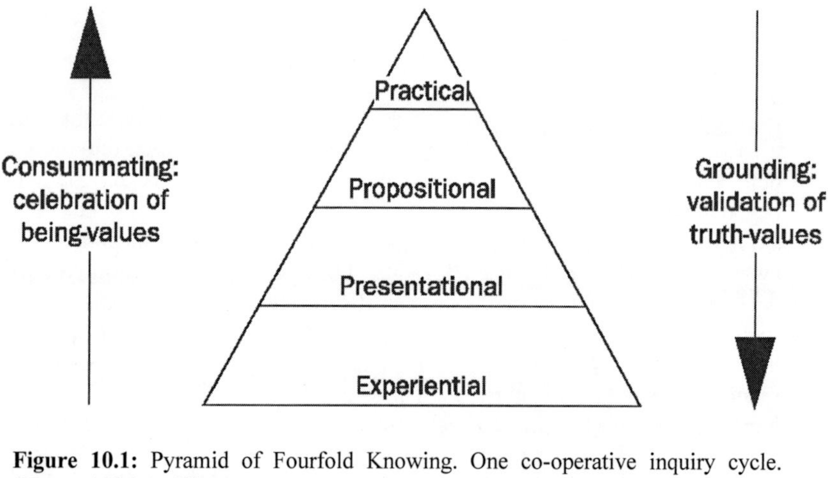

Figure 10.1: Pyramid of Fourfold Knowing. One co-operative inquiry cycle. (Heron, 1996, p. 53)

In some research methods, the roles of researcher and subject are mutually exclusive: the researcher only contributes the thinking that goes into the project, and the subjects only contribute the action to be studied (Heron & Reason, 2011). In co-operative inquiry these exclusive roles are replaced by mutual relationships, so all involved work together as both co-researchers and co-subjects. Everyone is engaged in the design and management of the inquiry; everyone gets into the experience and action that is being explored; everyone is involved in making sense and drawing conclusions; everyone engages in all cycles of action and reflection. One of the innovations of co-operative inquiry is the attention to forms of knowing that traditional research may be less interested in, such as experiential, artistic/creative and practical forms of knowing (Riley & Reason, in press).

There are two action and two reflection phases in co-operative inquiry. *Phase one* is reflection on the focus and nature of the inquiry. It is to agree on the inquiry question, clarify the inquiry topic, plan the first action phase, and decide on the recording methods for experiences. Phase two is action. It is to engage whole heartedly in the experience as agreed, inquiring into the question in an intentional way. Phase three is to build upon and deepen the experience in embodied, creative, presentational, or propositional ways. During phase four, co-researchers meet and reflect upon the original questions. They may decide to continue the same inquiry or amend the questions and direction for the next inquiry cycle. Here, the first cycle ends. When the second inquiry cycle begins, phase one is less naïve than in the earlier cycle, because there is experience, creativity, newly developed propositions, and embodied practice to build upon (Reason, 2002). In the approach we advocate, this cycling may continue–for life, perhaps, to enable the acquisition and practice of new skills and ways of knowing, being, and doing.

Whilst co-operative inquiry has its original roots in humanistic psychology (Wicks et al., 2008) our recent work has extended this to include other-than-human persons. Co-operative inquiry has several features, through which human communities can develop relational knowing with the more-than-human world. These are: it treats those involved–both human and other-than-human persons–as subjective, autonomous beings who are potentially equivalent participants in the inquiry process; it gives emphasis to the experiential ground of knowing as foundation for the poetics of presentation; and it affirms that practice is primary (Kurio & Reason, 2022). Whereas it may be said that much research has a propositional bias, co-operative inquiry has a bias toward practical knowing as the consummation of inquiry in worthwhile action in the world. Co-operative inquiry is guided by propositional knowing, inspired by presentational forms of knowing, and grounded in and continually reinvigorated through experiential encounter (Heron & Reason, 2011). Our emphasis in the present chapter is on experiential knowing, with its attention to deepening and widening our attention to phenomena, because the urgent task is to re-center relational knowledges and the agency of the more-than-human.

First and second person experiential inquiry

Having introduced living cosmos panpsychism and the co-operative inquiry method, we now show how first- and second-person experiential inquiry can be drawn on, to bring this perspective into everyday life. We begin with the notion of living life as inquiry, a first-person action research for life (Marshall, 2016). This involves a person taking up an inquiry approach to all they think, do, and feel. It is an embodied and reflective process of being curious about one's perspectives, assumptions, and actions; and is dynamic and critical. It involves an aspiration to ongoing learning, keeping in mind such properties of dynamic systems as connectedness, the tenacity of patterns, resilience, and emergence. Marshall (2016) emphasises systems thinking, recognising that parts cannot change without some shift in the systemic pattern, while acknowledging that parts can influence change in the system. It is this stance we bring to this co-operative inquiry, as first-person researchers sharing and reflecting upon our experiences of a sentient world.

We link this process of 'first-person' inquiry to the 'second-person' practice of co-operative inquiry (Reason & Bradbury, 2008) which provides a context for each person's investigations. Second-person inquiry describes action research approaches with two or more people inquiring together, having an aspiration to form a community for investigating questions of mutual concern (Marshall, 2016). The process we use integrates first person inquiring into the individual experience of engaging with their river and returning to the group to discuss and share findings with others who have also engaged with their rivers.

In the inquiries we have initiated over the past three years, primarily through the short course program at Schumacher College in Devon, England, we have invited human participants, in collaboration with rivers worldwide, to join us in addressing the questions we identified above: *What would it be like to live in a world of sentient beings rather than inert objects? How would we relate to such a world? And if we invoke such a world of sentient presence, calling to other-than-human beings as persons, might we elicit a response? Implicit in these questions is another: How might we listen to rivers, and learn to relate in ways that reveal ancient knowledges often occluded by the mechanisms of the West?*

At the time of writing, these inquiries have involved around fifty participant co-inquirers in relation to rivers across the planet. As initiating facilitators, we invite participants to start from the following inquiry protocol:

- We will visit our Rivers, exploring how to invoke presence and attend for response at least once a week.
- We will compose accounts of our experience (prose/poetry/drawing/photos), and share with the group for feedback and discussion.
- We will meet on Zoom weekly to reflect on the previous week's encounter with rivers and decide how we might proceed the following week.
- We commit to engaging in this way for six weeks, with a possibility for further engagement in the future.

Thus, our second person river inquiries start from the assumptions of living cosmos panpsychism. We engage in cycles of inquiry: visiting our rivers regularly to explore different approaches to invocation; finding initial presentational form in writing, photography, video, poetry, and drawing; meeting weekly to share experiences and make sense together; and deciding the practical actions we should take in the subsequent cycle (Kurio & Reason, 2022). Finally, joining together in collective reflection we do our best to make sense and draw out themes.

The co-operative inquiry groups create a context for first-person inquiry, opening for both encouragement and critique. They also provide a wealth of experience from which we intend in time to draw some general conclusions as to how modern humans can open themselves to a sentient world. In the present writing we limit ourselves to vignettes drawn from our personal journals. We have deliberately chosen excerpts that show the ordinariness of this work: we want to draw attention to the specialness in the everyday.

Our daily experience is to walk along or sit beside rivers or wetlands. There are two ways in which this is a second person inquiry, which involves people co-inquiring into questions of mutual concern (Marshall, 2016). The first way is that our learnings are later shared with human co-inquirers to conceptualise and more deeply understand experience. The second way is to interpret 'people' as persons of all species. As explanation, a philosophy of living cosmos panpsychism sees all beings as co-subjects within a communicative order rather than separated objects in an inert, unfeeling world. As personal response, this calls for an ecological or reconnecting sense of self that is inherently enmeshed in a field of relations (Mathews, 2021). Meaningfulness in this metaphysic is a quality of mind-nature, where mind is implicated in the field of relations, without privilege or separation. There is no possessiveness here, only mutual selves which is, in the end, a moral position at the heart of the cosmos (Mathews, 2021). This is second person inquiry.

On living life as inquiry in England: Peter's reflection on practice

March 7, 2022. (Very early spring.) Notes from River.

Here is a sample of journal writing. I regularly rise before daylight and drive then walk to a little sitting place beside some fallen branches, near the confluencc of the Frome and Avon rivers. My practice is to visit and sit with rivers, listening, observing, participating. I do this a couple of times per week.

Lately, I've been giving thought to ceremony to show respect to the living presence of River. I've been reflecting on qualities like grace, following my book called *In search of grace: An ecological pilgrimage*. The book records my trip sailing a small yacht called Coral around the British Isles.

> It's so quiet. No human noises. Standing on the bridge I catch the glint of the water and the bubbling of the stream under the bridge. The sky is already brightening in the east with the first hints of dawn; in the west it is dark still but with the glow of light from the town. The cockerel keeps calling, he definitely thinks it's morning! I go through the kiss gate into the field; when I get to the middle I stand in the field and just listen; it is almost silent. There's the owl calling, a background hiss of traffic, and the drone of a plane high above. No breeze, everything very, very still.
>
> As I walk across the field, crows begin to call, just solitary voices at first as if waking up. I notice the glint of the Frome River to my left through the trees. As I get closer to my spot at the confluence of the two rivers, I feel the world opens, there's a spaciousness. I stand at the base of the little peninsula with the Avon to my right, skeletal trees reflecting in the smooth water. There is the glimmer of the morning light catching on the water; a shimmer that takes my attention. I reflect on the concept of shimmer, described as 'ancestral power' by Australian environmental humanities academic Deborah Bird Rose (Bird Rose, 2017). There is so much going on here, and much we have to learn. I am learning to pay attention.
>
> I bow to Rivers. Even though I have the audio recorder going and so have a certain self-consciousness, I hold the intent to let go of my sense of separate self and be part of the wider whole. "Good morning, Rivers, Good morning. Here I am again. Thank you for being here. Thank you for your teachings over these years". I call the powers of the Four Directions. I call the power of the East, I call for new illuminations and insights, for new beginnings. I call the power of the West. I call for deep intuition, for an experiential knowing that deepens my sense of living in a sentient world, moment by moment. I ask that I learn that deep sense of continuity, being part of the living cosmos. I call to the south. I call for the children. I call the power of the north. I call for wisdom. I call for the wisdom of the four-footed ones. I

call for wisdom for our culture at this terrible and ridiculous time. Blessed be. What else is there to say?

I sit on the bank opposite the willow tree reflected in the dark silver of the river. The birds waken around me: tweets of wrens, tits, the caw of crows, the quack of a duck, the honk of a pheasant, now and again an owl. A thrush seems to be calling out 'do you agree? Do you agree?' Now there is a line of pale orange along the horizon; it's much lighter. I soften my gaze and attend to the birdsong. I find this leads me to feel less separate.

On living life as inquiry in Australia: Sandra's reflection on practice

November 5, 2022. (Spring). Notes from River walk: Evening.

Here is an excerpt from a piece of journal writing. I regularly visit the Leschenault Estuary and Collie River, Wardandi Country, South West Australia.

I am walking along the Leschenault Estuary on wet sand, where the high watermark meets dry sand and salt tolerant vegetation. I feel at home here. I sing out to N*gangungudditj walgu*[5] (Northover, 2008) the hairy faced snake, and tell him it is me, Sandra. I mention that when I sing out in that way, I feel *Estuary* 'opening out'–breathing–and starting to engage, becoming alive, becoming responsive. I accept this is my response–because *Estuary* is always awake, breathing with tidal rhythms and responsivities. However, I get out of my business-as-usual mind frame, and I respond. I feel beckoned to walk in the water. Evening is a stunning time of day when the place is full of energy. I feel the lure, the magnetism of Estuary. I cannot participate with Estuary without feeling 'alive', without feeling a zing in my step. I passed the casuarina trees and continued to the confluence of the Leschenault Estuary and the Collie River near Bar Island. I sing out again, and in an otherwise slightly wavy, very lightly wind-reactive surface, I notice beside me a glassy, shimmering, radiant, and openly welcoming patch of River. Transfixed, I breathe in the sacredness of River while I absorb its majesty. After several minutes of silent awe, I continued my walk.

I turn east and am walking alongside Collie River now. Here I feel that River calls me to family attentiveness and response-ability. I feel this is my heartland. I sing out to River again, and–as if in direct response–a school of

[5] I give respect to the Wardandi Noongar people of this area–past, present and emerging. I acknowledge the Noongar Elders and teachers who have taught me Noongar language and Stories over the years, with whom I have worked and learned. I thank you for this privilege.

> white bait leap out of the water, flashing and shimmering in the evening sun. While walking, I use imagination to visualise a Noongar (Indigenous) sense of time as pastpresence, in this Wardandi Noongar Country. I walk where people live together with more-than-human worlds within the river environs. This is a long now, where past and present fuse in [seasonal] time, so that as I look around, I understand that all that has been here is still present as species or in spirit, shadow, or out of sight. I visualise myself relating to all things, within nature. I later reflected that I was seeing many more connections than I usually do. I also noticed these perceptive abilities stay with me–this is a way of upskilling.

In this Story, I visualise a Noongar sense of time, part of the Noongar cultural worldview of this long-peopled place (Poelina et al., 2020), to enhance recognition, to 'see' more. We are all response-able–that is, we have an ability to respond (Bawaka Country et al., 2019; Northover, 2008). This is my daily practice, usually recording notes, photographs, or video clips. I walk co-operatively with other-than-human beings while reconnecting through River communications. I feel embodied knowing through experience.

Shared reflection–Peter and Sandra

As mentioned above, these vignettes make no grandiose claim; rather, they are ordinary, everyday. We invite, we sing out, we feel-with, we watch with sensitivity and care, we learn to love more-than-human beings and places, and we recognise the need to *un*learn no-longer useful binaries and other untruths. We invoke, we hold small ceremonies, we respect the conatus of the beings we meet, and in the respectful spirit of reciprocity, we respond. We become familiar with places–we notice the presence of birds, animals, breezes and winds, and the ways each engages with the other. We draw on our capacity for Imagination, in the sense that Blake used it to open perception to a world of transformed sensory experience; where identity is Imagination as world (Billingsley, 2018). This enables envisioning and recognition of our places and their animate worlds as ecological family (Poelina et al., 2020). We notice sentience everywhere, for sentience and presence surrounds us. We become deeply familiar with places we live with–as relational learners and knowers, recognising ourselves as self-in-relation.

As a practice, at the core of co-operative inquiry is a recognition and abandonment of hierarchical relationships, transforming these to authentic co-operation evidenced by, for example, an experience of the joy and wonder as self-in-world that is steeped in relational identity. The essence is a sense of connection among self and earth others, such that maintenance of the local living cosmos is the overarching purpose of existence.

Methodological implications

Research that negates the value and primacy of experience and practice, causes researchers and ordinary people to learn to distrust their ordinary observations and everyday experiences. As a system of power, through its assertion of worldview and positionality, traditional research removes people's capacity to engage fully in, or make sense of experience (White et al., 2018). Alternatively, research can enrich experience with perception, such that the visible, tangible, and observed interpenetrate feelings/experiences/ imaginal realms and critical/conceptual qualities of reality. This is an important substrate for co-operative inquiry.

We have shown in this chapter that over centuries, the orientation towards theoretical understanding that philosophers and researchers of the West have used, has repressed the communicative, responsive order of the world that surrounds us. It has also, through monopolising knowledge in the hands of the elite, repressed Westerners' own capacities for creating experiential knowing of the world around them. Thus, many have long denied that when we sing out to River, we may be graced with a response. And yet these phenomena are all around us–in rivers, the countryside, and cities–in our faces and our hearts, accessible through our ordinary senses. We are not arguing for the diminution of theories, but we are saying that it is only part of a story: theories can limit, as well as expand, our perception. For flourishing and renewal for all, the everyday empirical beckons our presence, noticing and participation. In the inquiry process, as we engage together and support each other in paying attention to a living world, that world becomes alive to us—partly because we notice more, and partly because the world itself becomes more responsive.

Once we understand we all are part of an intelligent, evolving, and communicative cosmos, we can be open to the whistles, songs, thumps, rustles, and patterns of more-than-human beings seeking reactions from us. We do exist in such a world–sentience surrounds us–and we are called on to relate in a way that is respectful and responsive. Ancient invitations such as these may have been unnoticed by researchers in the West through our scientific materialist history, which has entrenched dualistic thinking into the everyday habits of modernity (Mathews, 2017b). But now things are changing fast. Climate carnage and obliteration are with us, and we know there are new but ancient ways to relate with more-than-human worlds of beings. We wonder what other voices have been rendered mute by thinking by itself, rather than thinking-with embodied experience. There is much more to the world–a world of perception and experience able to reveal long-occluded knowledges that demonstrate the world may not be as we think.

Educational implications

We encourage teachers and children to respond to the beckons and calls of nature, recognising that liveliness is everywhere. For more on this, see works such as Blaise and Hamm (2020), Rooney et al. (2021) and Wintoneak and Blaise (2022). These authors allow children to respond to the communicative order of places, rather than teachers directing attention to things and giving instructions. Many Indigenous people see Land as a first teacher (Tuhiwai Smith et al., 2019). This is a different position to that of teacher-centred methods of instruction. For these types of experiential and creative/visual/imaginative learning, teachers and students need to leave the classroom, perhaps visiting playgrounds, rivers, wetlands, parks, and beyond to participate in, observe, and practice human responsive capacities over the seasons.

We recommend teachers draw on co-operative inquiry to invite children into cyclical experiences of action and reflection for learning; and to encourage children to learn 'everyday co-operative inquiry' as young as possible. By this, we mean enabling students to *discover* how to perceive: how to feel, smell, see and hear; and to acquire multiple creative ways to record and present evidence of a communicative cosmological order. It is not to direct children's attention though or tell them what teachers see. The relationships we ignore are visible, discernible, and practical, and invitations abound. Using the writings of Mathews (2021), it seems many of us in the West have *un*learned how to live. Young children are probably more responsive to worldly phenomena, and they have less to *un*learn.

Once teachers take up co-operative inquiry, we foresee classrooms everywhere being thriving places for learning, participation, creativity, arts, love of life, caring, noticing, relational healing, and transformation. Allowing children to genuinely participate in their reality is to offer hope for a damaged planet, hope for some form of flourishing to return from the ashes of narrow-minded thinking.

References

Abram, D. (1996). *The spell of the sensuous: Perception and language in a more-than-human world*. Pantheon Books.

Bawaka Country, Suchet-Pearson, S., Wright, S., Lloyd, K., Tofa, M., Sweeney, J., Burarrwanga, L., Ganambarr, R., Ganambarr-Stubbs, M., Ganambarr, B., & Maymuru, D. (2019). Goŋ Gurtha: Enacting response-abilities as situated co-becoming. *Environment and Planning D: Society and Space*, *37*(4), 682-702. https://doi.org/10.1177/0263775818799749

Berry, T. (1988). *The dream of the Earth*. Sierra Club.

Billingsley, N. (2018). *The visionary art of William Blake: Christianity, romanticism and the pictorial imagination* (1st ed.). I.B.Tauris & Co. Ltd. https://doi.org/10.5040/9781350989375

Bird Rose, D. (2017). Shimmer: When all you love is being trashed. In A. Tsing, H. Swanson, E. Gan, & N. Bubandt (Eds.), *Arts of living on a damaged planet* (pp. 51-63). University of Minnesota Press. https://doi.org/10.5749/j.ctt1qft070.7

Blaise, M., & Hamm, C. (2020). Lively Emu dialogues: Activating feminist common worlding pedagogies. *Pedagogy, Culture & Society*, 1-17. https://doi.org/10.1080/14681366.2020.1817137

Cresswell, I., Janke, T., & Johnston, E. (2021). Australia state of the environment 2021: Overview. Independent report to the Australian Government Minister for the Environment. https://soe.dcceew.gov.au/sites/default/files/2022-07/soe2021-overview.pdf

CSIRO. (2022). *State of the climate*. Commonwealth. http://www.bom.gov.au/state-of-the-climate/2022/documents/2022-state-of-the-climate-web.pdf

Díaz, S., Settele, J., Brondízio, E., Ngo, H. T., Guèze, M., Agard, J., Arneth, A., Balvanera, P., Brauman, K., Butchart, S., Chan, K., Garibaldi, L., Ichii, K., Liu, J., Subramanian, S. M., Midgley, G., Miloslavich, P., Molnár, Z., Obura, D., . . . Zayas, C. (2019). *Summary for policymakers of the global assessment report on biodiversity and ecosystem services – unedited advance version*. U. IPBES: International Panel on Biodiversity and Ecosystem Services, UNESCO, FAO, UNDP. https://www.ipbes.net/system/tdf/spm_global_unedited_advance.pdf?file=1&type=node&id=35245

Ghosh, A. (2021). *The Nutmeg's curse: Parables for a planet in crisis*. University of Chicago Press.

Harding, S. P. (2009). *Animate earth*. Green Books.

Harding, S. (2022). *Gaia alchemy: The reuniting of science, psyche and soul*. Inner Traditions Bear and Company.

Heron, J. (1996). *Co-operative inquiry: research into the human condition*. Sage Publications.

Heron, J., & Reason, P. (2011). Extending epistemology within a co-operative inquiry. In P. Reason & H. Bradbury (Eds.), *Handbook of action research* (pp. 366 - 380). Sage Publications.

Kurio, J., & Reason, P. (2021). Voicing Rivers through ontopoetics: A co-operative inquiry. *River Research and Applications*, *n/a*(n/a). https://doi.org/https://doi.org/10.1002/rra.3817

Kurio, J., & Reason, P. (2022). Voicing Rivers through ontopoetics: A co-operative inquiry [https://doi.org/10.1002/rra.3817]. *River Research and Applications*, *38*(3), 376-384. https://doi.org/https://doi.org/10.1002/rra.3817

Marshall, J. (2001). Self-reflective inquiry practices. In P. Reason & H. Bradbury (Eds.), *Handbook of action research: Participatory inquiry and practice* (pp. 433-439). Sage.

Marshall, J. (2016). *First person action research: Living life as inquiry*. SAGE.

Mathews, F. (2003). *For love of matter: A contemporary panpsychism*. SUNY Press.

Mathews, F. (2009). The world hidden within the world: A conversation on ontopoetics. *PAN: Philosophy Activism Nature,* (6), 97-109.

Mathews, F. (2017a). Come with old Khayyam and leave the wise to talk. *Worldviews*, *21*(3), 218-234. https://doi.org/10.1163/15685357-02103002

Mathews, F. (2017b). The dilemma of dualism. In S. MacGregor (Ed.), *Routledge handbook of gender and environment*. Routledge.

Mathews, F. (2017c). Invoking the real: from the Specular to the ontopoetic. In A. Gare & W. Hudson (Eds.), *For a new naturalism*. Telos Press.

Mathews, F. (2019). Living cosmos panpsychism. In W. Seager (Ed.), *The Routledge handbook of panpsychism*. Routledge.

Mathews, F. (2021). Law in the living cosmos: The 'ought' at the core of the 'is'. In J. Farris & B. P. Göcke (Eds.), *The Routledge handbook of idealism and immaterialism* (pp. 481-495). Taylor & Francis Group. https://doi.org/10.4324/9781003202851-39

Mathews, F. (2023). *The Dao of civilization: A letter to China*. Anthem Press. https://anthempress.com/on-the-dao-of-civilization-a-philosopher-s-letter-to-the-supreme-leader-pb

Mueller, M. L. (2017). *Being salmon, being human: Encountering the wild in us and us in the wild*. Chelsea Green Publishing.

Northover, J. (2008). *Joe Northover talks about Minningup Pool on the Collie River*. South West Aboriginal Land and Sea Council. https://www.noongarculture.org.au/joe-northover-minningup-pool/

Plumwood, V. (1993). *Feminism and the mastery of nature*. Routledge.

Poelina, A., Wooltorton, S., Blaise, M., Aniere, C. L., Horwitz, P., White, P. J., & Muecke, S. (2022). Regeneration time: Ancient wisdom for planetary wellbeing. *Australian Journal of Environmental Education*, 1-18. https://doi.org/https://doi.org/10.1017/aee.2021.34

Poelina, A., Wooltorton, S., Harben, S., Collard, L., Horwitz, P., & Palmer, D. (2020). Feeling and hearing Country. *PAN: Philosophy Activism Nature,* (15), 6-15. http://panjournal.net/issues/15

Reason, P. (2002). Editorial introduction: The practice of co-operative inquiry. *Systemic Practice and Action Research*, *15*(3), 169. https://doi.org/10.1023/A:1016300523441

Reason, P. (2017). *In search of Grace: An ecological pilgrimmage*. Earth Books.

Reason, P., & Bradbury, H. (2008). Introduction. In P. Reason & H. Bradbury (Eds.), *Sage handbook of action research: Participative inquiry and practice* (pp. 1-10). Sage Publications.

Reason, P., & Heron, J. (1997). A short guide to cooperative inquiry. http://www.human-inquiry.com/cishortg.htm

Riley, S., & Reason, P. (in press). Co-operative Inquiry. In J. Smith (Ed.), *Qualitative psychology* (4th ed.). Sage Publications.

Rooney, T., Blaise, M., & Royds, F. (2021). With shadows, dust and mud: Activating weathering-with pedagogies in early childhood education. *Contemporary Issues in Early Childhood*, *22*(2), 109-123. https://doi.org/10.1177/1463949120939202

Tsing, A. L. (2021). *The mushroom at the end of the world: On the possibility of life in capitalist ruins*. Princeton University Press.

Tuhiwai Smith, L., Tuck, E., & Yang, K. W. (2019). *Indigenous and decolonizing studies in education: Mapping the long view*. Routledge.

United Nations. (2022). Pakistan's 'climate carnage beyond imagination', UN chief tells General Assembly. *UN News: Global perspective Human stories*. https://news.un.org/en/story/2022/10/1129337

Van Dooren, T. (2022). *A world in a shell: Snail stories for a time of extinctions*. MIT Press.

White, P., Wooltorton, S., & Palmer, M. (2018). Confronting, collaborating and crafting: An enlivening methodology for academic ecojustice activism. In S. Garvis & A. L. Black (Eds.), *Women activating agency in academia: Metaphors, manifestos and memoir* (pp. 111-122). Routledge Ltd. https://doi.org/10.4324/9781315147451

Wicks, P. G., Reason, P., & Bradbury, H. (2008). The SAGE handbook of action research (2nd ed.). SAGE Publications Ltd. https://doi.org/10.4135/9781848607934

Wintoneak, V., & Blaise, M. (2022). Voicing Derbarl Yerrigan as a feminist anti-colonial methodology. *River Research and Applications*, *38*(3), 435-442. https://doi.org/https://doi.org/10.1002/rra.3822

Sandra Wooltorton (BA (Hons), BEd, MEd (Hons), PhD) - University of Notre Dame Australia, Nulungu Research Institute, 88 Guy Street, Broome, Australia.

Sandra Wooltorton is Professor and Senior Research Fellow with the Nulungu Research Institute, University of Notre Dame Australia, and an Adjunct with the Centre for People, Place and Planet at Edith Cowan University. She is interested in the relationship between people and place; and is an advocate and campaigner for social and environmental justice. Sandra conducts research in the environmental humanities and education. She is concerned with the dynamics of intergenerational cultural transfer in a more-than-human world beyond Anthropocene scenarios; and is currently engaged in a series of experiential and co-operative inquiries exploring living cosmos panpsychism in relation to rivers.
https://orcid.org/0000-000108677-870X

Peter Reason (BA (Hons), MA, PhD) - University of Bath, School of Management, Bath, United Kingdom.

Peter Reason is an international leader in the development of participative approaches to action research. Since retiring from full time academic work, he has focused on writing books and articles that link the tradition of nature writing with the ecological crisis of our times drawing on scientific, ecological, philosophical, and spiritual sources. He undertook two ecological pilgrimages at sea; explored the place of art and poetry in a time of ecological catastrophe; writes and sings 'songs for the Earth'; and is currently engaged in a series of experiential and co-operative inquiries exploring living cosmos panpsychism in relation to Rivers.

CHAPTER 11

ATTUNING WITH MORE-THAN-HUMAN KIN: EXPERIMENTING WITH WAYS TO DE-PRIVILEGE HUMANS WHEN REFIGURING EDUCATION IN UNCERTAIN FUTURES

PETA J. WHITE, JO RAPHAEL, SHELLEY HANNIGAN & ROBIN BELLINGHAM

Abstract

Our chapter describes an unfolding inquiry with coinciding practices that explore complexities and affordances for entanglements to reveal educational opportunities that mirror the urgent need for change in the ways humans impact Earth and kin. We are a collective of education academics enacting regenerative education that experimented with multi-species thriving to re-focus pedagogical practices in these challenging times. We wanted to explore and enable the sharing of our 'attuning with' more-than-human-others. The desire to collaborate with colleagues in intentional and inclusive ways led to the improvisation and adaptation of a Padlet (Map of the world). We then collectively unpacked how these attempts at 'attuning with' could, and perhaps should, bring insight to and changes in our education practices. A focus of our 'attuning with' was to de-privilege the human and to employ the agency of more-than-human voices. 'Attuning with' requires listening with more than ears, seeing with more than eyes, and connecting with hearts, and includes kin of any kind. In this chapter, we showcase our exploration mediated via Padlet, connect with colleagues who are not similarly located, prioritise inclusivity of more-than-human voices, and reflect on resulting adaptations to our pedagogical practices.

Key words: Anthropocene, Multi-species thriving, Pedagogy, Experimenting, Assemblages

Prelude: Education in challenging times

The naming of the Anthropocene as an epoch (geological era) representing human induced and significant changes to Earth's systems (IPCC, 2023) should be acknowledged as alarming. All sectors must enact urgent strategies for change across the globe, especially in the Global North and especially education. The 6th synthesis report from the IPCC (Intergovernmental Panel on Climate Change) warns that the "pace and scale of climate action are insufficient to tackle climate change" (2023, np). Educators are well positioned to initiate and support societal redesigns and reimaginings for sustainable futures (UNESCO, 2023). This chapter reflects on the work of educators who are engaged in generating new insights and practices that intend to lead to establishing new ways of being in our world–sustainable ways of being. We, the authors, are acutely aware of the human induced challenges faced by all and look to learn, enact, and re-inscribe more attuned ways going forward.

Our growing polycrisis forms the uncomfortable backdrop for this methodological exploration into pedagogical practices for regenerative futures. Human-induced climate change since colonisation and the industrial revolution (Lewis & Maslin, 2015), biodiversity loss with the 6th mass extinction (Dirzo et al., 2014) felt strongly across Australia, and the privileging of inappropriate economic models based on unlimited growth in finite systems which result in the environment continually subsidising to the point of destruction (Raworth, 2017; Donut Economics Action Lab, 2023).

We named our collective of academics 'Regenerating Education for Uncertain Futures' and aspire to theorise and refine practices that prioritise the inclusivity of humans and beyond. Throughout 2022 we met online and in-person, when possible, to explore and develop our methodology and practices. This chapter captures a representation of our methodological experimentation and exploration through our process. We felt compelled to respond and feel our ways forward in this inquiry, as opposed to applying 'ready-made' methodological practices. We challenged our practices to develop pedagogies so that ecological systems and kin are prioritised, relationships valued, and we begin to decolonise our methodologies and practices of education. Our experiments began in 2022 and the implications and theorising of our enactments has been reflective praxis (carried into 2023).

In this chapter, we report on our enactments of experiments to deprivilege humans in pedagogy and to 'attune-with' more-than-human kin. Our Padlet activity facilitated a collaborative mapping of material and situated relations with more-than-human kin. Following this, we focus our attention on the affective intensities, materiality, relations, and stories emerging with the spaces, places, beings, and matter we have mapped. We reflect on this mapped assemblage as a partial enactment/emergence of our own becoming-with-the-world. That is, we consider our interconnectedness with place (Malpas, 2018) and that "the universe is not simply a place but a story–a story in which we are immersed, to which we belong, and out of which we arose" (Swimme & Tucker, 2011, p. 2). Matter in every form is a co-produced meaning and embodiment of the world, or "storied matter" (Oppermann, 2020, p. 411), in which we ourselves are entangled (Barad, 2007; Riley & White, 2020).

Working in collaboration was a process we enjoyed (White et al., 2020), applying arts-based methods to self-study of teacher education methodology was now familiar (Hannigan et al., 2016), similarly exploring our practice through autoethnography was also comfortable (Hannigan et al., 2022; Raphael et al., 2022). Individually, we have been broadening into collaborative inquiries (Wooltorton et al., 2021) and more recently (Sutton et al., in press) as strategy to include kin in our explorations.

Opening: Multi-species thriving and learning with kin

As university educators we are familiar with and increasingly concerned about discourses of education dominated by linear, abstracted, and supposedly objective ways of learning about and conceiving of the world, in which the supposed best educative practices are technicist, mechanistic, and instrumentalist. These notions of good education have developed via the entangled foundations of imperialism, enlightenment thinking, colonialism, capitalism, and scientism in the thinking and habits of being of modernity. Among other concerns, this thinking entrenches divisions between humans and nature and the idea that knowledge is a means to "index, control, and order the world" (Stein, 2019, p. 3). Success is equated with wealth and resource accumulation. In education and pedagogy these logics and values have been accepted as part of a universally relevant epistemology that offers "certainty, predictability, and consensus" (Stein, 2019, p. 3). The contemporary education system is therefore fed by a sense of entitlement to authority, mastery, certainty, and universalism (Stein et al.,

2021), producing abstracted, disembodied, dislocated knowledge and supposedly singular and essential truths.

These views in education help to reproduce and reinforce the conditions in society by which the rest of the (non-human) universe is understood to be a non-agentic, largely non-sentient resource, and to rationalise and justify ongoing colonising and exploitative practices and ways of thinking for production, ongoing growth, and accumulation (Ferreira da Silva, 2016). The resulting eco-social fractures are increasingly impossible to ignore.

The epistemologies, practices, and implications which circumscribe our workplaces trouble us significantly as they do not reflect our own educational values, aspirations, and deep commitments. However, being mainstream, they do to some extent reflect our own educational backgrounds and training. In this chapter, rather than focusing our attention and energies on critical analysis of this situation in education, and/or crafting the logic explaining better ways of approaching pedagogy, we instead work to disturb these influential ways of learning and being and to understand, learn, and relate otherwise.

Our diverse experiments include inquiries that partially and/or temporarily dissolve boundaries of learner/teacher, place/space, ecology/self. They attend to how space, place, matter, language, histories, and pedagogy shape both the possibilities and constraints of what we can know. We consider how supposed 'others' in our mapped places always merge with ourselves; how it is that we know these 'others' through ourselves (Rautio, 2017) and ourselves through 'others'; and how pedagogical agency and possibilities emerge through these entangled relationships with 'others'. These attuning experiments include reflections on how the 'self' is "always already multiply dispersed and diffracted throughout spacetime(mattering)...in its ongoing being-becoming" (Barad, 2014, pp. 181-182). That is, our identities and selves are not singular, separate, and determinately bounded, but are differentiated and dispersed in multiple, hybrid and distributed, material and virtual forms in the world (Barad, 2007), in and across selves / waterways / mountain ranges / Zoom classrooms / studio spaces / dogs / aliens / friends / lace curtains / silhouettes / sharks.

'Attuning with'… the purpose of the practice

Our research program, Regenerating Education for Uncertain Futures, began at a time when we needed to build a research community and reconnect after an extended period of being disembodied and dispersed, having met mostly via screens during a pandemic. In building our research community we recognised that existing within the neoliberal university had us acting more as solitary isolated trees, rather than a forest capable of collecting and sustaining ourselves through mycorrhizal networks (Davies, 2021). Neoliberalism, Davies (2021) explains, "sets us against each other, separating us from the earth others with whom we are intimately enfolded" (p. 317). We needed to commune, coalesce, and gel together, and we wanted to do this within the natural environment and with our more-than-human kin. We sought a creative approach to research that could provide new fertile ground as a basis for our methodological experimenting and for being/becoming academic together.

Our first practical challenge was to find ways to engage our colleagues in a creative approach to research that involved collaborative located and embodied theorising, as an antidote to the pandemic period of disconnection and disembodiment from which we were emerging. This sounds relatively achievable, yet we were located across three campuses across the state of Victoria, Australia, and some colleagues were travelling internationally. Additionally, mobility in the outdoors was challenging for some of us, and we needed to find ways to enable all our colleagues to participate. Beyond this we wanted to provide opportunities that were inclusive and that provided ample time to experience and reflect.

Collaborative cartography

We proceeded to build an inclusive community through a process of collaborative map-making using an online mapping tool on the web platform Padlet[1] (Figure 11.1). We chose a map template that offered a satellite view of the world, showing the features of the land, waterways, oceans, and the built environment. We decided on a collaborative cartography strategy to value place-kin relationships in their complexity. The affordances of using a Padlet world map included the possibility of searching a location and dropping a pin, with the geographic coordinates automatically added. Images, sound recordings, videos, web links, text, and documents can be added to each pinpoint. It is possible to view the map as the entire world and then zoom in to a precise location.

---

[1] Padlet (https://padlet.com/) is an online platform that enables participation from multiple authors simultaneously. Many platforms exist that are similar, Padlet was chosen for the 'world map' template.

Figure 11.1: The Padlet template for collating our experiences with many of the pins representing attunement experiments. The second image is the content of one pin–a photo of a tree ecosystem and a reflection describing the 'attuning with' experiment.

Beginning at a School of Education conference session, we invited colleagues to walk together, literally, or figuratively, as we de-privileged the human and employed the agency of attunement to more-than-human voices embedded in our digital map. We explained that attunement requires listening with more than ears, seeing with more than eyes, and connecting with hearts. We created space for experimentation as we considered how more-than-human voices guide us as educators. Colleagues who were present at the conference session were invited to find a partner or go alone, to find space, outdoors if they chose, and to add a pin or 'voice' to the map. We then reconvened, shared the voices, and asked, what is needed now, what is needed next, and how might this attunement enable us to teach more inclusively?

Colleagues who were unable to attend in person, and even those who were present, continued over time to develop their own practices of attuning and pin more-than-human voices to our collective map. Pins were both local and spread across hemispheres and continents. The more-than-human voices ranged from forests to flowers, silkworms to signs, beehives to bacteria, fungi to frogs, and artworks to air. Each pin to the map expressed an intensity, an attuning with and a learning encounter. Our map, although digital, was both the Earth (mapping the world) and earth (locating our place-kin connections), it was a fertile common ground, and although we did not always share the same time or space, we walked on this metaphoric ground together, sharing stories.

Entanglements of 'attuning with'

Presented below are four entanglements drawn into a linear representation as examples of our experiments of 'attuning with' the more-than-human within our exercise of collaborative cartography. Each author offers a brief description of their experiment in text and with visual representations that were pinned on our collaborative cartography. We use the voice or 'I' and 'we' without differentiating us as individuals or specific authors. We have worked with each other in many experiments, but not all. Our individuality is not what matters here–it is the experiment, the intent, the methodological and pedagogical challenges and resolutions.

Entanglement 1–Decentering the classroom

Our educational methodology and practice regularly enable entanglements with the spaces and places beyond the classroom. We generate ways to engage with the materiality of our ecosystems and affirm our places within our interconnected, complex, and changing systems. We move around the inattention that has resulted in human induced climate change leaving us stranded and uncertain amidst our self-generated polycrisis. I appreciate collaborating with my colleagues–they like to read and think and do and think. We have nurtured in our practice a strong methodology and pedagogy of wonder and engagement with the more-than-human world with an ethic of multi-species thriving, and care for self and others. This experiment reflects on learning opportunities we enjoyed with secondary teachers as we co-designed and developed methodological practices and generated pedagogical opportunities for use with their (future) students.

I have had the opportunity to work with secondary teachers a lot lately and I noticed that often they focus on practical pedagogies to efficiently enable conceptual learning and, as a consequence, apply language that has the tendency to re-inscribe exploitative industrial practices. The embodied learning possibilities are often ignored with preference given to inside, book focused pedagogy.

When invited to consider the challenges of the Anthropocene and human-induced ecological crises, they seemed unsure. In my world, these were commonly used notions that are entangled with deep grief and are uncomfortable, yet I was somewhat surprised by our teaching colleagues' unfamiliarity. I unpacked it with them and felt their energy and spirits plummet, but not their enthusiasm. They reported that they shared my grief and now had additional ways to explain and explore these feelings. We discussed how our shared work gave them permission to continue to explore through less conventional ways, more embodied ways, how to enable their young people (students) to come along with them as we learned and changed.

Learning in community

We planned to meet outside, determined to proceed despite the weather, however, we were fortunate to have struck a perfect spring day–the first warmer day we had experienced this year. Our educational task was to explore 'earth systems' and to gather pedagogies that enabled student learning, encouraged interest and engagement, and enacted new priorities for learning with the environment and not just about it.

My hopes were exceeded as we spoke with elders about their learning experiences and knowledges, explored communities of more-than-human kin (Figure 11.2), and formed an educational assemblage without impact as we canoed over banjo sharks, leaving them undisturbed and us profoundly connected. We collaborated to build our knowledge about these ecosystems through interactions, explorations, wonderings, and observations. We collected data and valued naming to understand relationships, but we prioritised questioning reasoning to textbooks and field guides. We heard stories of colonisers' adventures as they learned about their new landscapes. We appreciated localised knowledges of our Aboriginal elders as they lived with the land–owned by the land and sea. The shift in language and prioritisation from owning the land and sea was well heard by all.

Figure 11.2: Kin learning communities

The single day suggested methodological and pedagogical opportunities to how we might practice learning with and about our environments, place-based (McKenzie, 2005) and embodied (Dewey, 1929). Value was attributed to creating learning entanglements where assemblages were unravelled as well as enjoyed in their infinite complexity (Wooltorton et al., 2021). We appreciated the opportunity to share knowledges and ways of being with each other and beings beyond usual classrooms. We valued slow pedagogy (Payne & Wattchow, 2009) encouraging time for aesthetic and appreciative practices. Perhaps we gave permission for us all to continue to look for ways to generate learning assemblages outside the classroom and with our more-than-human kin.

Entanglement 2–Metaphoric meanings of waterways

During 30 years of practice as an artist, I have been fascinated how, despite an auto-ethnographic focus in my work, the bodies that might represent self or kin are something else; more-than-human. I have come to understand that the birds, trees, water forms, dresses and other symbolic imagery that re-occur in my art over the years, have come to represent something far greater than self alone. In my world, water re-occurs. I understand it as an interconnectedness we have with our world; we are after all mostly made of water ourselves. I created a film during the Covid pandemic and shared this with other images on the Padlet map. Seeing these pins of my own place-experiences on our world map, I reflected on how film had captured my relationship with flows and interruptions during the pandemic years: the partial and moving reflections of (for example) lace curtains on internal walls which pick up the silhouette of my dog (who accompanied me during lockdowns where I taught my classes on Zoom). This was contrasted with my filming at wild places such as national parks I visited with waterfalls, creeks, rivers, and the sea, once lockdowns ended. I had montaged this footage together with stops and starts representing the pivoting modes of education and art practice we had experienced during this time (see Hannigan, 2021).

This focus on water (Figures 11.3) also caused me to reflect on research I conducted with other environmentalists concerned by the waterways between the Otway National Park and Geelong, Victoria (Bartier et al., 2020). I reflect further back on a body of artwork I created in my first year at art school in Christchurch, New-Zealand over three decades ago. Struck by the visibility of the snow-capped mountains nearby and needing to cross a small river to get from my home to my university studio, I developed a series of prints in the late 1980's that explored this journey of the river intertwined with my new self-place (Malpas, 2018). Now, still in an academic institution, but in another country, Australia, water separates and connects my two home countries of New Zealand and Australia that are displayed on the Padlet map. I share this re-occurring exploration of water in my research/practice to remind us that we all probably have these levels of knowing that possibly get overlooked in the interests of haste, modernity, workplaces etc. Slowing down and engaging in our creative projects, contemplative thinking and being able to find these deep and re-occurring metaphors and symbols through these experiences, might remind people that they are there, and I think important to attune with.

Figure 11.3: Upper Glenelg River, Balmoral, Victoria

Ingold (2013a; 2013b) and London (2003) suggest ways of investigating creativity of artistic practice as creativity in place and creativity of place. I realise how both demonstrate relationships with the more-than-human. I encourage students to explore their own place and identity interconnections in and with the more-than-human, in holistic and embodied ways. I share my own and other artists approaches to doing this while they engage in making and learn about mediums, materials, tools, styles, genres, and visual forms.

Engaging in my water piece, highlighted how the more-than-human can also apply to art that may on the face of it seem to be about the artist's self. As artists tend to explore their experiences of place (Dear, 2011; Sullivan 2005), the inseparability of self-place (Malpas, 2018) emerges in their practice in a myriad of totemic, symbolic, and metaphoric ways. This value of art practice needs to be promoted more to shift art education away from conventional notions of it being mere picture making, aesthetic, or decorative. To achieve this, we need to show examples of how artists and educators work in embodied, place-based ways, and out of their experiences meaningful insights emerge and are presented and communicated.

Entanglement 3–The studio as a space of becoming

As our collective mapping grew, I noticed the outdoors and natural world was well represented. For me, there is an indoor space that I have attuned with over time. It is the space of the university drama and dance studio in which I often teach drama and dance education for pre-service teachers. It is neither the walls nor other physical fixtures that define the space, but rather the space / air / atmosphere between them. I thought, if air and art can be pinned to our more-than-human map, then so too can the space of the studio, as more-than-human kin. It is a space that holds many years and layers of pedagogical encounters, and through attuning with the studio space over time, I have discovered an ontology of learning and teaching. As a way of sharing this with my education colleagues, I pinned it to the map.

Although familiar to me, the studio differs from the standard university teaching spaces that are typically filled with an orienting configuration of tables and chairs, often in rows, facing a screen and a tutor's table or lectern (Figure 11.4a). I have often wrestled with such furniture trying to turn a standard classroom into a space suitable for the kinds of activities that drama and dance education involves.

I had the opportunity to work with architects to design the drama and dance studio in a program of building refurbishment. It is a large spacious room, smooth grey carpet, and black curtains from ceiling to floor, to absorb the sound and light when required (Figure 11.4b). There is an absence of furniture, with chairs and rostra blocks quietly stacked and concealed behind curtains, to be used only if needed. I have walked into the studio thousands of times by now, and every time I feel a sense of surprise and excitement about the next class and the possibilities for learning that it holds. I have an affinity with this room, my teaching companion, and I feel we work well together.

Figure 11.4: Contrasting spaces for learning– (a) the classroom and the (b) studio

I located the precise spot on the map and pinned a 30 second video of the space, allowing the viewer to move through the anteroom along the corridor and through the door with its sign 'Drama & Dance Studio'. They are taken inside the room, to observe from different angles. Interspersed throughout the video are one-second, almost subliminal flashes of students engaged in activity, as haunting moments. Finally, the view is of the empty space. The space speaks:

I am E2.014, a drama studio, an open uncluttered furniture-free smooth space of possibility.
I am an empty space.
I am an invitation and a question.
I ask, what possibilities can a space like me open for learning?

My work with pre-service teachers, both generalist and specialist in the arts, means I observe many bodies interacting within this space. When entering the studio for the first time, students, accustomed to standard classrooms, instantly recognise that this space is unfamiliar and demands different modes of learning and teaching. At the level of sensation, the space causes affects and 'percepts', feelings, and perceptions in a plane of immanence (Deleuze & Guattari, 1994). I discern a flash of delight in some. Eyebrows raised, a small smile, they stride in, anticipating something different, something new.

Amongst other students, I perceive a flicker of concern across faces. Eyes dart, shoulders hunch, they are disorientated, uncertain of where they should place themselves, or what they should do with their bodies. Uncomfortable in an unfamiliar space, they cling to the edges, perhaps wondering: Where is the front? Where are the tables to hide behind? Where are the chairs that will shape my body into sitting, and position me safely in relation to others?

Space is not neutral or passive.
Space is active in causing sensations.
The studio is a space of becoming.

Deleuze and Guattari (1987) theorise space as smooth and striated, and while the studio is inevitably both at times, the clear open area offers a blank space, a lack of certainty, and a question. While students (and teachers) may experience different intensities of comfort or discomfort when they first encounter it, the studio space provides opportunities for bodies to interact differently (Figure 11.5). Bodies in and with space move in and out of assemblages, with changing qualities, lines, flows, speeds, and intensities. Three actors embody a concept. Four puppeteers animate a paper horse. Six dancers achieve unity in motion. Twenty students/teachers sit in a circle to yarn. Paying close attention to the studio space has awakened me to its agency and potential. The sense of surprise experienced in the first moments of entering causes a small disruption, creating an opening for learning. This is a space where learners and teachers can be transformed in endless ways– multiple assemblages, and new relationships for future world-making. Together we learn about teaching and learning both within and from our studio space.

Figure 11.5: Assemblages and becomings in the studio

Entanglement 4–Co-enabled becomings with the Tararua Ranges

Below is the beginning of a short story I wrote which was inspired by time spent in the Tararua Ranges (Figure 11.6) near to where I lived in Aotearoa New Zealand. I pinned this image of the Tararuas on the Padlet map and am taken back to the special localised green-gold quality of the light on/with the hills at evening and morning.

Figure 11.6: View from Mt Holdsworth Tararua Range New Zealand by Geoff Jacobsen and Lisia Grocott. Licensed under CC BY-SA 3.0.

Out beyond the end of the unsealed road at the north end of town there's a track winding up and over two ridges and up that track there used to be some abandoned wooden cabins, some kind of spiritual retreat in its day. I headed up there one time, I had known them to be empty for a long time and I had hit on an idea of sitting on a porch and doing a little practice out to the valley.

I picked a likely one and got in the door and there he was inside, lying in the corner, looking over at me with yellow old eyes. He was tucked up on a woven flax mat on the dirty floor, rolled in grey blankets. The cabin had just one room with wooden plank shelving, a small chair with hard straight arms, nothing else. He lay there on the floor, under the shelves.

He needed help, I could see that. I went out there every day and fed him and gave him water.

Sometimes I wasn't sure if my friend was very young, like a newborn, or very old. He was creased and black, all black. I have never seen such a thin wizened thing. Loose greasy skin on his face, limbs, neck; all deeply creviced. His eyes were swollen and they looked out as if they had looked too far and were weary of it. His look was as ancient as they come, and he was as helpless as a bubba. (Bellingham, 2010, p.106)

I am interested in how place, materiality and writing together co-enable possibilities; how they facilitate material-discursive becomings. At the time, I wrote: "For becoming we need a more imaginative, deeper, and more passionate kind of instinct for truth and freedom" (Bellingham, 2010, p. 6) than logical reasoning can offer, and this can be enacted in fiction. 'Story' can be understood as a relational node between place, materiality, self and writing, through which new agency can emerge (Haraway, 2016). Through this story emerged an interest in the kind of becoming that might result from a potent relationship with a particular place and a creature so strange that it defamiliarises our assumption that we can ever know others (places, beings, friends) in any objective, deterministic sense, and so the relationship works to engage and connect on other terms. It is from this opening up of relations that possibilities come, not from any objective knowledge of 'the other' we might believe we hold. In SF (science fiction, speculative fiction and fantasy) writing the figure of the alien can support this kind of defamiliarisation (Gomel, 2012), estranging us from the familiar and making the familiar feel strange. This story helped me experience some release from the ways that we entrench divisions between ourselves and the matter and beings who we see as 'not us', or 'other', and simultaneously, to wonder about the other/the alien within the self (Bellingham, 2010).

The story also facilitated expressions of intensities of one-ness with place, and of grief, loss, aporia and resignation to loss, through the weary alien. Re-experiencing/reimagining this place/creature and creating this story played a role in feeling my way toward a position in relation to planetary degeneration and irreparable damage, which is something to do with attuning with and holding compassion for the weariness of the planet (including ourselves) for all that it has been subjected to, and also the necessity of accepting death and decay as essential phases in ongoing planetary relations (including, and perhaps especially, the death and decay of our human selves, human-kind, and our current ways of being). In terms of an emergent pedagogy this experiment suggests the potential value of place-based writing, poetics, and practices of defamiliarisation drawn together to bring to life or to regenerate intensities of material, sensory, being, and relationship. Also, the value and fun of playing around with selves and with relations as multiple, distributed, and emergent across beings, places, and landforms.

What now, what next, and how?

We ask ourselves: what is needed now, what is needed next, and how might 'attuning with' practices enable us to teach more inclusively, with careful focus towards our more-than-human kin. The world map on the Padlet, provided a topological way for the participants to literally pin images from their methodological and pedagogical experiences, to share with others. These images were anonymous, and each represented a snippet or moment of time from their experiences, that were shared on the Padlet gallery, open to the interpretation of others. The technology of the Padlet offered extended time for us to go in and add images and visit others' contributions. In doing so we created a collaborative cartography gallery (Figure 11.1), a myriad of experiences that the four authors have taken time to delve into further, to understand each other's reasons for pinning these images and what the images represented.

Our emergent methodology and pedagogy of attuning with the more-than-human, works to encourage and support each other and other learners to wonder and inquire with being alive, and with the infinite complexity and paradoxes presented by more-than-human kin and with places. As others have also found (Riley & White, 2020; Sutton et al., in press), in our experiments here we find that generating and noticing aesthetic and affective intensities in experiences is an important process in this regard. Our experiments in this chapter generate and attend to compelling entanglements involving particular and situated encounters with light, shade, mountains, coasts, waterways, trees, animals, classrooms, furniture, grief, aporia, weariness, stress, discomfort, culpability, anticipation, delight, enchantment, stillness, flow, bodily communications, movement, connectedness, interdependence, closeness, warmth, intrigue, beauty, and wonder. The pedagogical and ethical value of this approach is to affirm the need to take seriously and consider closely that these experiments, as part of encounters with distributed forms and places, are communications from/with "a meaningfully articulate planet" (Oppermann, 2020, p. 413), and to make efforts to interpret the narratives they signal.

Many Indigenous peoples are already well prepared to listen to and interpret such signals from Country (Povinelli, 2016). In this writing we work with various experimental means of connecting with / listening to / expressing / interpreting place-based signals, including through touch and senses, play, immersion, movement, and creative writing. These actions to generate, notice, express, and interpret aesthetic and affective intensities with more-than-human kin help to decentre human narratives, to strengthen respect for more-than-human expressions and agency, and to grow our capacity for much needed learning from and with them.

Our experiments in experiential re-interpretation also indicate, as many others have too, that disruption of learning expectations is generative in facilitating more-than-human wonder and inquiry. In this chapter these disruptions often involve place and space. Moving ourselves to new places such as local coastlines, creating unexpected learning spaces such as open studios, and defamiliarising thinking about what places of learning are or should look like, for example by focusing on learning in waterways, imagined places, or anywhere where vibrant, challenging, or dynamic life asserts itself, often generated a heightened sense of anticipation, loosened boundaries, and animated a more open range of learning possibilities. As teachers and learners, we look for ongoing opportunities to share these kinds of learning disruptions and displacements in virtual, fictional, material, and dynamic settings.

We also reflect on the fact that the cutting together-apart (Barad, 2014) that we do in the experiments and stories created here co-constitutes different intensities and marks on ourselves and the world, in ways that support the actualisation of different versions of ourselves and the world, for which we have co-accountability and co-responsibility. Considering this, we feel the weight of importance of the forms of methodology and pedagogy we choose, explore and support. We consider how to facilitate creative, critical, and response-able co-storying of coevolution and future world-making. There are many, many forms that this could take in methodology and pedagogy. One of the ways that we have explored this in this small set of experiments is in attempts to enact and express selves as distributed and co-emergent. These are in part, ethico-political moves, as they enact alternatives to and escapes from the dominant representation of beings in the world as determinately separate, objectively bounded, self-interested, competitive selves. This ontology naturalising individualism and rivalry is a major obstacle to conceiving of ourselves as connected to and co-responsible for the rest of the planet, and therefore is a massive limitation for our responses, in solidarity, to planetary, social, and ecological crisis. We have experimented with expressions of ourselves distributed across phenomena in the world in waterways, arts and artefacts, alien beings, and our colleagues and learners, to assist us to wonder about where we begin and end, and how we, and all things, are connected and co-constituted in an ongoing sense and are therefore both constrained by co-constituting forces, as well as dynamic and full of potential (Barad, 2007).

In our methodology and pedagogy, we are conscious of the responsibility to learn about and support both creative exploration and critical understanding of the co-constitution of ourselves and more-than-humans as kin. We discuss with our pre-service teachers and co-learners that, counter to dominant political education discourses, individualism and competition are not the only ways to understand the fundamentals of being. Also, that there is no value-neutral or ideology-free education because the ways we teach and learn bring and strengthen knowledge forms and mattering stories in the world which have lasting material marks on and consequences for how agency, freedom, privilege, accumulation, inequity, damage, and exploitation, are patterned and distributed in the world.

Our pedagogy involves realistic confrontation with the fact that for us and most of our students it is not easy to let go of entrenched humanist, colonising logic, but also a continued commitment to explore with our learners our collective responsibility to understand, interrogate, and creatively transform what it is possible to know and ways of knowing. These are ongoing sites of inquiry in our methodology and pedagogy of attuning with more-than-human kin.

Through our research methodology we continue to explore four different approaches as educators / researchers and four possibilities for experiencing / teaching others / learning about self / kin and the more-than-human. Sharing this, in this paper offers other educator researchers' insights into this methodology of doing research together (and with a larger group) that is not restricted to a particular time and place (therefore inclusive). On the back of others who have explored embodied practices to focus on the more-than-human voices and challenge the traditional education systems (see Knight, 2020; Davies, 2021), our work contributes to normalising embodied practices as researchers / educators. This helps steer a movement away from the focus of self, ownership, disciplines, ego etc. that are not friends to the environment.

Towards matters with our worlds

> It matters what matters we use to think other matters with; it matters what stories we tell to tell other stories with; it matters what knots knot knots, what thoughts think thoughts, what descriptions describe descriptions, what ties tie ties. It matters what stories make worlds, what worlds make stories.
> —Donna J. Haraway (2016)

Our experimental collaborative cartography was a response to a desire to do research and being / becoming academic differently. We allowed ourselves to be vulnerable, as we each took risks in posting our chosen personal encounters, knowing our post may reveal more about us that we can even know ourselves. Some of our mapped encounters are presented as carefully composed images that are aesthetically satisfying, such as a majestic mountain or the coast from the air. Others are quick snaps, common, curious, incomplete and imperfect, but there is a kind of beauty in them all. Winston (2010) reminds us of the significance of beauty in learning. He proposes beauty as a concept that can provide a necessary counter to utilitarian values, such outcome driven curriculum, arguing that beauty in education provides for the mysterious and uncertain, and affords hope and pleasure. There is beauty in our collective map, and beauty, it would seem, is an imperative as we seek healing, a transformative approach to research, and pedagogies for uncertain times.

Maps are made to help us understand our place in the world. Our collaborative map-making has placed us on the map and connected us to Earth / earth like the mycorrhizal networks beneath a forest of trees. As we moved between each other's mappings, we became more connected not only to each other, but to our more-than-human kin. At the beginning, we had wondered how more-than-human voices may guide us as educators. We have become attuned to them, listened to their wisdom, and realised that rivers, rooms, frogs, bees, post-boxes, hen houses, banjo sharks and birdsong, to name a few, can indeed be powerful co-educators.

As educators with multiple practice fields and creative tendencies, we have highlighted pedagogical affordances from our experiences. We have done this within a supportive research group that has enabled us to explore methodological affordances with each other. With teaching and learning and creating a better world / environment our core concerns, we have shared our research and our way of doing research to show how methodology is pedagogical but also how pedagogy can be methodological.

Acknowledgements

We acknowledge our Country, traditional owners and Elders, colleagues, time to un/re-learn afforded by our privileged positions in academia. We acknowledge the Regenerating Education for Uncertain Futures research program in the School of Education at Deakin University, and all our colleagues who have participated in our collective cartography. We acknowledge our focused efforts to discuss, learn, practice methodology and pedagogy differently and with a focus on our more-than-human kin. We recognise the interconnection of self / place which means we aren't acting alone.

References

Barad, K. (2007). *Meeting the universe halfway: Quantum physics and the entanglement of matter and meaning*. Durham: NC: Duke University Press.

Barad, K. (2014). Diffracting diffraction: cutting together-apart. *Parallax, 20*, 168-187.

Bartier, J., Hannigan, S., Gardiner, M., & Mathison, S. (2020). Embodiment of Values, *IDEA Journal, 17*(2),180-200.

Bellingham, R. (2010). *Friend in the woods* (unpublished Masters thesis), Auckland University of Technology, Auckland, New Zealand.

Davies, B. (2021). Moving beyond (id)entities, toward emergent becomings of the world and its mattering. *Australian Journal of Environmental Education, 38*, 311-327.

Dear, M. (2011). Creativity and place. In M. Dear, J. Ketchum, S. Luria, & D. Richardson (Eds.), *Geohumanities: Art history, text, at the edge of place* (pp. 9-18). Oxford, UK: Taylor & Francis

Deleuze, G., & Guattari, F. (1987). A thousand plateaus: Capitalism and schizophrenia. London and New York: Continuum.

Deleuze, G., & Guattari, F. (1994). What is philosophy? London: Verso.

Dewey, J. (1929). *Quest for certainty*. New York: Minton, Balch and Company.

Dirzo, R., Young, H. S., Galetti, M., Ceballos, G., Isaac, N. J., & Collen, B. (2014). Defaunation in the Anthropocene. *Science, 345*(6195), 401-406.

Donut Economics Action Lab. (2023). *About doughnut economics.* https://doughnuteconomics.org/about-doughnut-economics

Ferreira da Silva, D. (2016). On difference without separability. In J. Volz, J. Rebouças, I. Rjeille, P. Fundação Bienal de São, & M. Pavilhão Ciccillo (Eds.), *Incerteza viva: 32nd Bienal de São Paulo: 7 Sept-11 Dec Catalogue* (pp. 57-65). https://biennial.com/files/pdfs/7879/d-ferreira-da-silva-on-difference-without-separability.pdf.

Gomel, E. (2012). Posthuman voices: Alien infestation and the poetics of subjectivity. *Science Fiction Studies, 2*, 177. doi:10.5621/sciefictstud.39.2.0177

Hannigan, S (2021). SALVEage film. https://www.shelleyhannigan.com/

Hannigan, S., Raphael, J., & White, P. J. (2022). Evolving teacher education practice through collaborative arts-based autoethnography. In E. A.

Anteliz, D. L Mulligan, & P. A. Danaher (Eds.), *The Routledge international handbook of autoethnography in educational research.* Routledge. https://www.routledge.com/The-Routledge-International-Handbook-of-Autoethnography-in-Educational/Anteliz-Mulligan-Danaher/p/book/9781032119922

Hannigan, S., Raphael, J., White, P., Bragg, L., & Cripps Clark, J. (2016). Collaborative reflective experiences and practice in education explored through self-study and arts-based research. *Creative Approaches to Research, 9(*1), 84-110.

Haraway, D. (2016). *Staying with the trouble: Making kin in the Chthulucene*. Durham, NC: Duke University Press.

Ingold, T. (2013a). *The conical lodge at the centre of the earth-sky world.* Architecture Lecture Series. http://progressivegeographies.com/2013/06/16/placing-lecture-tim-ingold/ Portland State

Ingold, T. (2013b). *Making*. London, UK: Routledge.

IPCC. (2023). *Synthesis report of the IPCC sixth assessment report (AR6).* https://www.ipcc.ch/report/ar6/syr/

Knight, L. (2020). Inefficiently mapping boundaries: How is an urban citizen. *Mapping Meaning the Journal*, *4*, 66-75.

Lewis, S. L., & Maslin, M. A. (2015). Defining the Anthropocene. *Nature*, *519*(7542), 171-180.

London, P. (2003). *Drawing closer to nature: Making art in dialogue with the natural world (1st ed.).* Boston, MA: Shamballa Publications.

Malpas, J. (2018). *Place and Experience (2nd ed*), Routledge.

McKenzie, M. (2005). Second thoughts on post-critical inquiry. *Environmental Education Research*, *11*(4), 455-462.

Oppermann, S. (2020). Storied matter. In R. Braidotti & M. Hlavajova (Eds.), *Posthuman glossary* (pp. 411-414). London, UK: Bloomsbury Academic.

Payne, P., & Wattchow, B. (2009). Phenomenological deconstruction, slow pedagogy, and the corporeal turn in wild environmental/outdoor education. *Canadian Journal of Environmental Education, 14,* 15-32.

Povinelli, E. (2016). Geontologies: A requiem to late liberalism. Durham, NC: Duke University Press.

Raphael, J., Hannigan, S., & White, P.J. (2022). Understanding pedagogies for climate change education through collaborative arts-based autoethnography In P.J. White, R. Tytler, J. Ferguson, & J. Cripps Clark (Eds.), *Methodological approaches to STEM education research volume 3*. Cambridge Scholars Publishing.

Rautio, P. (2017). "A super wild story": Shared human–pigeon lives and the questions they beg. *Qualitative Inquiry, 23*(9), 722-731.

Raworth, K. (2017). *Doughnut economics: Seven ways to think like a 21st century economist.* Penguin

Riley, K., & White, P.J. (2020). Attuning-with' affect, and assemblages of relations in a transdisciplinary environmental education. *Australian Journal of Environmental Education, 35(*3), 262-272.

Stein, S. (2019). The ethical and ecological limits of sustainability: A decolonial approach to climate change in higher education. *Australian Journal of Environmental Education, 35*(3), 198-212.

Stein, S., Andreotti, V., Hunt, D., & Ahenakew, C. (2021). Complexities and challenges of decolonizing higher education: Lessons from Canada. In S. Kumalo (Ed.), *Decolonisation as democratisation: Global insights into the South Africa experience* (pp. 48-65). South Africa: UKZN Press.

Sullivan, G. (2005). *Art practice as research: Inquiry in the visual arts.* Sage.

Sutton, B. A., Bellingham, R. A., & White P.J. (in press). Deepening our capacity for teaching with place. *Australian Journal of Environmental Education 39*(3).

Swimme, B. T., & Tucker, M. E. (2011). *Journey of the universe*. New Haven, CT: Yale University Press.

UNESCO. (2023). *What you need to know about education for sustainable development.* https://www.unesco.org/en/education-sustainable-development/need-know

White, P., Raphael, J., Hannigan, S., & Cripps Clark, J. (2020). Entangling our thinking and practice: A model for collaboration in teacher education. *Australian Journal of Teacher Education, 45*(8), 93-110.

Winston, J. (2010). *Beauty in education*. New York: Routledge.

Wooltorton, S., White, P., Palmer, M., & Collard, L. (2021). Learning cycles: Enriching ways of knowing place. *Australian Journal of Environmental Education, 37*(1), 1-18.

Peta J. White (BSc (Hons), MRSysMan, PhD, SFHEA) - Deakin University, School of Education, 221 Burwood Highway, Burwood, 3125, Victoria, Australia.

Peta J. White is an Associate Professor in science and environmental education at Deakin University. She educated in classrooms, coordinated programs, supported curriculum reform, and prepared teachers in several jurisdictions across Canada and Australia. Peta continues her commitment to initial teacher education leading courses, units and programs, and in-

service teacher education through research-informed professional learning programs. Peta's current research follows three narratives: science and biology education; sustainability, environmental, and climate change education; and collaborative/activist methodology and research.
https://orcid.org/0000-0002-0225-5934

Jo Raphael (BEd, MEd, PhD, SFHEA) - Deakin University, School of Education, 221 Burwood Highway, Burwood, 3125, Victoria, Australia.

Jo Raphael is Senior Lecturer in arts education at Deakin University. She is actively involved in teaching and researching in community and education settings within Australia and internationally. Her areas of research and publication include applied drama and theatre, inclusive education, and teacher education. In her practice she applies drama as a pedagogy for learning in areas across the curriculum including science, humanities, and education for sustainability and climate justice. Jo's research draws on arts-based and participatory action research methods that are inclusive of participants as co-researchers and collaborative arts-based autoethnography as part of a commitment to developing a community of practice amongst teacher education colleagues.
https://orcid.org/0000-0001-5220-314X

Shelley Hannigan (BFA, MEd, GradDipEd, Phd) - Deakin University, School of Education, Pigdons Road, Waurn Ponds, 3216, Victoria, Australia.

Shelley Hannigan is a Senior Lecturer in Art Education. Her specialist areas are visual and media art and wellbeing. She is a practicing artist and has a background in working as an art therapist in hospitals as well as teaching art in a range of education contexts. Found objects, collage, mixed media, and textiles have been introduced to her practice over the years as she engages materiality, embodiment, and place as both artist and academic. She researches in education, place, identity, practice, arts health, STEAM, and arts discipline areas and supervises Masters and PhD students in these areas.
https://orcid.org/0000-0002-3232-5761

Robin Bellingham (B Appl Sc (Health), Dip Teach, M HSc (Hons), MCW, PhD) - Deakin University, School of Education, 221 Burwood Highway, Burwood, 3125, Victoria, Australia.

Robin Bellingham is a Senior Lecturer in Pedagogy and Curriculum at Deakin University, Melbourne. Her research explores how education and methodologies can respond ethically to pressing problems of modernity such as educational and political disempowerment and disengagement, the ongoing effects of colonisation, and ecological crisis. She draws on different writing genres, posthumanism, new materialism, decolonial theory, and critical theory.
https://orcid.org/0000-0002-0425-6770

CHAPTER 12

THE PARTISANSHIP AND PERFORMATIVITY OF CREATIVE PRACTICE RESEARCH FOR ENVIRONMENTAL EDUCATION: QUESTIONS AROUND EDITING INTERVIEWS AND KNOWLEDGE PRODUCTION

CASSANDRA TYTLER

Abstract

An embrace of posthumanism in environmental education (EE) has supported grounded and embodied modes of practice-based research as a methodology. Emphasis has moved from representational knowledge to a celebration of affective encounters in place. This chapter draws on an art project undertaken at the Queen Victoria Market as a place-based methodology. I question the notion of agency of the artist/researcher in EE research and their neutrality in the process of presenting their creative research. Specifically, I look at audio interviews that I edited into a creative work and exhibited at the market as an audio-visual soundscape. If creative research is performative and takes part in processes of worlding, where does the neutrality in research lie? From a posthuman perspective, how can the EE single-subject-researcher presenting creative work remove their partisanship? To connect with these methodological questions, I look to Mazzei's theorisation of a *Voice without Organs* (2013) alongside Alaimo's theorisation of trans-corporeality (2008). I turn to Barad (2003) to argue the performativity of creative research challenges the authority conceded to language to define what is real. This is a step towards a pedagogy that opens itself beyond human exceptionalism to further learning about the human and the more-than-human.

Key words: Post qualitative, Post-humanism, Environmental education, Creative practice research

Storying our world

> Who told us we could, and should, separate our messy bodies from our pure minds lest they contaminate rational thought? When did we begin to believe our bodies were, indeed, absent from research and scholarship? (St. Pierre, 2015, p. 138)

> [T]he corporeal substance of the human is ultimately inseparable from 'the environment.' … '[N]ature' is always as close as one's own skin. (Alaimo & Hekman, 2008, p. 238)

Storytelling exists in many forms and is constructed in multiple ways, all of which can be powerfully productive. Stories *do* things, and they contribute to outcomes. Stories tell us, or teach us, or influence us on who we believe we are, what we deem 'natural', and how we think we should act. They can be cathartic, and they can be dangerous. Stories can have violent effects, especially when it comes to issues of race, gender, class, and nation. They shape us ideologically and bodily but can also be used as counter narratives to offset hegemonic identities. Often messages and affects embedded within a story depend on the storyteller, the mode of their delivery and their social positioning. As Haraway (2016) famously tells us:

> It matters what matters we use to think other matters with; it matters what stories we tell to tell other stories with; it matters what knots knot knots, what thoughts think thoughts, what descriptions describe descriptions, what ties tie ties. It matters what stories make worlds, what worlds make stories. (p.12)

Stories direct us but stories are not innocent. Their world-making potential produces knowledge through their iteration and citation: their performativity.

Butler (1999) has taught us the connection between the citational chain repeated and its material consequences on bodies through gender performativity. Here a reiterative naming and doing produces the gendered subject. Gender is not static, or a given, but through repetition that begins in nomenclature, becomes what is so often assumed innate. Accordingly, our bodies are materialised through regulatory 'stories' that are constantly reinforced over time. Our bodies are implicated into our sociality, and referring to the St. Pierre's quote above, are entangled with our research and knowledge production. In other words, our bodies matter, not only because they hold markers of social norms, but because they are materialised by and react to the world around them, our narratives or 'stories' included. In thinking about research practices and scholarship, the body can often be forgotten. Embodied responses that take place in creative practice research are a form of embodied learning, and a framework of scholarship that can create a rich understanding of our world in extended ways.

Referring to Alaimo's quote above alongside Barad's (2003) theory of intra-actions:

> The world is a dynamic process of intra-activity in the ongoing reconfiguring of locally determinate causal structures with determinate boundaries, properties, meanings, and patterns of marks on bodies. (p. 817)

The body, alongside all matter, which includes the non-human and more-than-human, is shaped and constantly changing in relation to their intra-actions. Barad extends Butler's conception of gender performativity to the performativity of all matter, encompassing the material and discursive elements that *make* bodies. Our world is in movement, adapting, reacting, and changing in reference to everything within it. There is an interconnection between embodied beings and the changing, dynamic material world. The human body is not a bounded vessel. We cannot claim to be disembodied or detached from the space we inhabit. We are influenced by the material and discursive as well as nature and culture. By extension, our stories, where they take place and how they are told, are embroiled within the material-discursive realm.

Evaluating knowledge

Directing attention to both scholarship and methodology then, especially that of environmental education (EE) that uses art practice, how are our stories storied to make worlds of knowledge and understanding that can also qualify within the academy? Is it necessary to counter one's implicit partisanship when making creative work that is also considered research within science education? Can creative outcomes be considered a valuable approach in EE when I as researcher am choosing what is included and excluded in the final creative outcome? I focus these questions in relation to a creative research project, *Frutti Tutti*, that uses interviews of participants in its outcome and will be discussed further below. I ask how this work can move beyond my singular subjectivity as interviewer and editor when I am the one making the choices as to what is heard and what is omitted. In erasing interviewees' voices the research is *composed*. It is edited. Therefore, where is the neutrality of the research? Can a story, in its broadest form, be included in STEME (Science, Technology, Engineering, Mathematics, Environment) education research? In going forward, I will map out the ways as well as reasons that indeed, creative practice work can be a valuable method and learning tool to discover and interact with ideas and concepts around EE. This involves a rejection of the Cartesian mind/body split, or logical positivism, still seen in much scientific inquiry now (see St. Pierre, 2015); a recognition of embodied learning as a powerful tool of knowledge creation; a de-centring of the human-singular-body-subject; and a focus on performativity beyond language and the agency born from it (Barad, 2003, 2007).

Instead of following a positivist method, I am inspired by St. Pierre's (2011) notion of post qualitative inquiry as well as critical posthumanism to decentre the humanist subject in my research practices. In this sense, from the outset I know that I am not separate from my research. I have an intention in making this creative work, and even though this aim may not be a finality, I am partial in my objectives. This is not a methodology, but I have various hopes with the outcomes, from the simple, 'I hope this is an engaging experience for participants', to the more complex, 'I hope that this work connects participants to each other and to the unbounded aspects of nature, all tied to their market-going experience'. In other words, I conceived of this creative practice work as a mode of learning through doing and experiencing in relation to the world outside the bounded human subject. I am not starting with a methodology, but do the preconceptions mentioned above create problems in claiming this work as research? I turn to Manning (2020) in her discussion of what she calls *research-creation*: "New processes will likely create new forms of knowledge that may have no means of evaluation within current disciplinary models" (p. 28). With this project I am thinking through how my creative research can be used within EE when the nature of a public-facing artwork is, through its very form, partial. Yet, this approach in research is also political. To quote St. Pierre (2015):

> The Cartesian knowing subject, that master of the universe, has destroyed the planet, and we now live in the age of the Anthropocene, the new geologic period marked by clear evidence that human activities have damaged the earth beyond repair. Our thirst for knowledge, our science for science's sake, our clever technologies—Science devoid of ethics—has produced despair and a longing for another way of being and living. (pp. 143-144)

Below I map out my approach to EE research and grapple with why the performative and its partiality do not diminish the research, and how creative practice can create modes of scholarship in new knowledge creation. I do this through an analysis of *Frutti Tutti,* an art project that takes the form of a walk where audio-visual works are mapped across the Queen Victoria Market in Naarm (Melbourne).

Frutti Tutti

My aim with the project, *Frutti Tutti*, is to reposition participants' self-recognition that they are a part of a larger ecosystem that they interact with in numerous ways; an ecosystem that is influenced by both nature and culture and all the politics that are entangled within this. To achieve this, I made a smartphone-based project that took place across the market space. For the first iteration, the video works were accessed via QR codes placed at different points on a map within the site. The intention was to performatively generate new connections as a way of imparting new forms of embodied knowledge.

I began this project as part of a three-week residency with Testing Grounds, which is housed alongside the Queen Victoria Market. I was given a large brick corner site, with sizeable windows, that was well-positioned to receive foot traffic. The sign on the window instructed passers-by to "come in and be interviewed about fruit". I recorded the interviews as sound files. The interviews were conversational. I had a set of questions that were not strictly adhered to but were driven by my own interest in the social, cultural, environmental, economic, and political effects that are produced in relation to fruit. I would allow participants to guide the discussion, so the interview topics were vast. As a starting point, I organised questions around a set of themes:

- Taste, texture, and memory
- Consumption and price
- Waste
- The environment
- Growing/production

I used the residency room as a hub to receive the public and I also moved around the market space interviewing growers and stall holders.

I made a soundscape of the conversations attached to a non-linear video. There were six videos in total that had fruit titles. On the exhibition day, I posted a map at my 'hub' that participants could follow in search of the QR codes. My idea was that they would listen as they walked amongst the market stalls of fruit, their senses piqued through the smells, sounds, and colours coming from the market space and at the same time hear different voices through their headphones. The goal was to have a multiplicity of voices, weaving in and out. *Frutti Tutti*'s aim was to immerse participants through the many 'voices', heard and listened to, technological and otherwise. The goal was that individual participants would become an embodied part of the market hub, the fruit they were surrounded by as they walked, as well as the interviewees who they would listen to while walking. They would be enmeshed in the constant materialisation of their worlding as they listened, moved, and watched. In interacting through these different means, they become more aware that 'nature' is not separate from themselves, and it is entangled in the stories that they are listening to and walking amongst.

In recording and editing *Frutti Tutti*, I set the questions and made decisions as to what segments to include and those to cut. The interviews went for a minimum of 15-20 minutes, but many were closer to 40 minutes. There was a lot of information to cut. By making those cuts, or agential cuts as Barad would call them (2011, p. 125)[1], how is the work impartial? I am making the decision as to what voice comes next in the soundscape, which can have the effect of acting as an 'answer' to a previous voice, either in agreement or disagreement. For example, one person said he didn't understand what organic fruit really was, and the next voice was of an organic fruit seller who explained the process. Soon after another person spoke about the cost of organics, and later someone else spoke about why he's prepared to pay more for organic fruit. In this sense, the interviews are not soundscape alone, they are informational and personal. Language, which includes tone, inflection and intonation, and the meaning derived from it cannot be escaped.

The voice without organs

To work through these questions, I turn to Lisa Mazzei's (2013) posthumanist work on the use of the voice in the interview to position my approach and address some of the concerns mentioned. Mazzei draws upon Deleuze and Guattari's *Body without Organs* (BwO), to theorise a *Voice without Organs* (VwO) as a way of thinking of voice as non-essentialist and not originating from a singular subject. The body of Deleuze and Guattari's

---

[1] Barad explains that an "intra-action enacts an *agential cut* … effecting a separation between 'subject' and 'object.' [It] enacts a 'local' resolution *within* the phenomenon of the inherent ontological indeterminacy". (p. 125)

(see 1984) *Body without Organs* is a fragmented and disassembled entity, rather than a coherent whole that is singularly articulated and unchanging. It is an assemblage of dynamic energies and forces. The body is potential rather than an individuated, hierarchical form that denotes a subject. To quote Deleuze and Guattari: "The BwO is opposed not to the organs but to that organization of the organs called the organism" (1984, p. 158). In other words, they trouble the idea of a clearly bounded, static, and closed form. This includes the human as well as sciences, philosophy, art, and society. Mazzei interrogates the interview voice in research from this unlocked perspective. She contends that to think of the voice as force rather than as a humanist subject bound to a singular body with a singular idea expressed through that voice, creates a presence that is always in a dynamic flow of becoming.

In this approach, the human is, to quote Mazzei (2013):

> …an assemblage, an entanglement, a knot of forces and intensities that operate on a plane of immanence and that produce a voice that does not emanate from a singular subject but is produced … in an enactment among researcher-data-participants-theory-analysis." (p. 733)

In other words, voice is not exceptional or separate from the situation in which it exists and does not originate from an isolated individual. Mazzei rejects a humanist qualitative inquiry that centres the interviewee subject. She questions the belief that the interviewee subject's account can truly epitomise a singular logical 'truth' about themselves and their situation. Further, the way voice is received and interpreted by the researcher is funnelled through their own experiences and unobjective interpretations. Mazzei instead thinks of the interview as an assemblage, alongside the BwO and VwO. She quotes Deleuze and Guattari's description of the assemblage:

> There is no longer a tripartite division between a field of reality (the world) and a field of representation (the book) and a field of subjectivity (the author). Rather, an assemblage establishes connections between certain multiplicities drawn from each of these orders. (qtd in Mazzei, 2013, p. 735)

Thinking of the interview in this way, fields of reality, which include the questions, the voice, the place, are not distinct, and are assembled alongside the representation of the interviewer's future research and the subjects involved in the endeavour. To quote: "There is no separate, individual person, no participant in an interview study to which a single voice can be linked–all are entangled" (Mazzei, 2013, p. 734). To understand a VwO is to approach human agency from a post-human position, most specifically through a reading of Barad's intra-actions. Agency does not exist in the singular subject but through material and discursive entanglements that are always emerging and changing in differing situations. It is through intra-actions that agency emerges, not through an all-knowing subject, nor one that is uniquely human. There is no hierarchy of knowledge. All knowledge is connected and co-constituted. Agency does not pre-exist the encounter and is mutually and emergently productive alongside the entanglements that do not sit within the one human entity. This post-human tactic dissolves speech spoken by the interviewee as the sole or exceptional authority of communication. Through the VwO, my edits of voice do not speak for the one organism–my own nor the person speaking.

Frutti Tutti's voice is not centred, but following Mazzei, is an assemblage of forces. My own voice is sometimes heard, not as authoritative questioner but usually 'voice as laugher' or 'voice as engagement'. I allowed my voice to be heard in the final work as a reminder that I am not separate from the commentary. As interviewees speak, they are not speaking as singular subjects. Meaning is derived through the *forces* and *intensities* that occur through the interview, where both myself and interviewee are becoming in new ontological entanglements (Mazzei, 2013, p. 738). I cannot explicate myself and my decisions in the presentation of the artwork. No one is a unique or essentialist subject, and following Mazzei, I too am entangled into the work. As a VwO, the voice, questions, edit, path, participants listening, their histories and backgrounds are all an assemblage of relational webs. I am not neutral and nor is the interview 'data' in making this research, even if I had written my interview findings up in an academic journal article rather than presenting them as a site-based artwork.

My decisions concerning what to edit in making the soundscape are guided by the entanglements I experienced while interviewing, my own personal connections before and after, and my positioning within the market space. I am an agent making decisions, but my agency comes from the assemblages: the VwO. In a similar way, the interviewees speak, not from the present moment of the interview alone, but through their own histories and experiences that form their stories of fruit. These are entangled in the material and discursive as well as the natural and cultural. For example, I immediately had flashes of childhood when a woman recounted that her mother would only allow her to eat mangoes in the bath. I don't know that this was true for me but listening I could feel the fleshy juiciness of mango in my little messy hands. A man originally from Indonesia accompanied by two family members spoke about durians and their repugnant smell that Caucasians would not be able to appreciate. His interlocutors laughed and sighed in what felt to me like glee at the diminished culinary tastes of Anglo-Europeans. My own feelings as a white settler in colonised, un-ceded land made me think of my own cultural heritage and disgust in my invader history. I enjoyed the teasing. A man from Nigeria spoke about the Igbo calendar, which is synchronised to the market days. He explained that time in many villages is measured around which village market is open. Learning this made we further aware of the controlling nature of temporality and its specific normalising organisation on our daily lives. Another interviewee spoke about pruning grape vines in a certain way so that when he picked the grapes later in the year, he wouldn't have to stoop too low making the physical labour less arduous. This made me think about work I would do as a younger person, never outdoors, but physically exhausting. I thought about the little tips and tricks one would pick up along the way. Then the romance of working outdoors made me think about French farmers working the fields and harvesting wine. I am sure this work is not as romantic as what I imagined, but my cultural longings brought me to this place. Here a VwO is produced that includes geography, ethnicity, culture (and cultural longing), labour, memory, politics, and the experiential. These aspects act on one another concurrently. They do not create a singular 'meaning' but relations are found amongst them. These relations come from my own attunement and the relationality created between the interviewees, the marketplace, fruit, and all the other 'organs' of the story-assemblage.

Impartiality is not the question alone here. What is at stake is ways that creative art research outcomes can be a tool for EE moving forward ethically within research in ways that open participants up to their place within the world and their environment. The goal is to create an embodied understanding of their connection to more-than-human worlds beyond their individual selves. This knowledge can be both cerebral and felt, as embodied knowledge. As Polyani (2009) states, "I shall reconsider human knowledge by starting from the fact that we can know more than we can tell" (p. 136). Once the artwork is experienced in the market, a new agency emerges–one that I am a part of but not central to. With the exhibited outcome of *Frutti Tutti* the participants are affectively entangled with the non-human, the organic, their phones, the material of the market space, the ground that they walk upon, as well as the conceptual, representational linguistic elements. This is a fluid assemblage across decentred bodies and thought (Deleuze & Guattari, 1984). I as artist and researcher am making agential cuts, through the editing and through planning the map participants follow, but I am not an agent in the humanist sense of being separate. Instead, even as I make these cuts, I am already always entangled through the intra-actions agentially occurring.

Language overemphasised

But what of the content being listened to or being omitted? While entanglement exists, the information from speaking is a part of the embodied knowledge being produced and I am controlling those decisions through the edit. However, as Barad reminds us, the representationalism of language is overemphasised. She asks, "[h]ow did language come to be more trustworthy than matter?" (2003, p. 801). Performativity, according to Barad, challenges the authority conceded to language to define what is real. As she contends, "[t]he move toward performative alternatives to representationalism shifts the focus from questions of correspondence between descriptions and reality (e.g., do they mirror nature or culture?) to matters of practices/doings/actions" (p. 802). Knowledge production through discursive practices is not formed through language alone but through multiple forms across various sites, steeped in the hegemonies of power. As Barad notes, "[d]iscourse is not what is said; it is that which constrains and enables what can be said" (p. 819). The representationalism of the linguistic as the dominant form of knowledge is problematised because it cannot escape regulation through power relations. A unified subject does not hold all knowledge. Instead, assertions and subjects "emerge from a field of possibilities" from active and varying inputs that are endlessly changing (ibid). It is through performative practices, where the doings are productive in new ways that disrupt the mind/body split, that varying and embodied knowledge relations can come into being.

Performative research is rooted in action. As described by John Law and John Urry, performative research methods "have effects; they make differences; they enact realities; and they can help to bring into being what they also discover" (2004, p. 393). This is not to say that realities are indiscriminate or random, but that they are epistemological and ontological, or in Baradian terms, onto-epistemo-logical. As Kat Jungnickel contends, "[e]very method furnishes different ways of seeing the world, and fundamentally shapes the findings" (2020, p. 14).

Trans-corporeality within a politics of location

So what does post-human performativity mean for educational scholarship? If we embrace the belief that we are connected to all things in a way that opens ethical entanglements and disrupts hierarchies between subjects and objects or between the human and non-human; if we reject the Cartesian mind/body split; then how do we impart knowledge and how do we measure its success? In responding to this question, the first step is beginning from a position that recognises and deconstructs the power imbalances and inequities that exist in certain humanist modes of knowledge production and from there interrogating what knowledge is outside of this system and how it can be 'knowable'. Central to this is the notion of the human subject. As Braidotti reminds us, "being human never was a neutral term. It never was an inclusive term" (CCCB, 2022, 1:19). Some humans are considered more human than others. Not all bodies are allowed to hold space in the same ways. Ideological apparatuses configure and fortify how bodies are seen and interacted with. As Ahmed (2008) aptly notes, "[w]hat bodies 'tend to do' are effects of histories rather than being originary" (p. 56). Naturecultures are emergent so that social and political systems, and histories cannot be considered separate from the materiality of the body and space. According to Alaimo, "[s]cience, medicine, history, law enforcement and the military are all entangled in the racism, environmental degradation and epistemological quandaries that course through [a] protagonist's very blood" (2018, p. 436). The invisibility of men, whiteness, the cis-gender, and the able-bodied within Eurocentric and American social and educational spaces must be accounted for when considering learning so that normative assumptions are not made around the affects for all participants as a monolith. Neimanis (2017) emphasises the need for a "politics of location" when considering embodiment:

> Where is my body? When is it? Why is it – that is, thanks to what, and whom? What are the membranes that separate it or differentiate it from others? Where and how do those membranes break down? Where and when does that body cease to be? And then: in what ways does it repeat? … [A]n understanding of one's body … is multiscalar and mutigenerational; porous and palimpsestic. It is a congeries of the personal and the political; of the material and the semiotic. It is biological and cultural, and it is never only one thing, in only one place, or only 'itself'. (p. 29)

Here, the material facets of pedagogy overlap with more-than discursive methods of knowledge production. While problematising a strict and singular meaning or outcome, this also positions research as a diffractive and adaptive endeavour.

A sense of the body singular or lived-experience singular is problematised when we try to do research from a post-humanist standpoint. How can we talk about 'lived experience' while also de-centring the subject? How do we not essentialise subjects in this way? We are intra-acting and responding according to our surrounds and histories with the human and more-than-human within our sphere. Knowing this and moving forward understanding the power relations that privilege particular bodies above others, combining the material, discursive and 'lived experience' of participants, is an ethical ontology of relations in research.

I turn to trans-corporeality alongside the BwO to de-centre the researcher and participants. Trans-corporeality is the neologism that Alaimo gives to describe the interconnection of embodied beings and the changing, dynamic material world. To quote: "By emphasizing the movement across bodies, trans-corporeality reveals the interchanges and interconnections between human corporeality and the more-than-human" (Alaimo & Hekman, 2008, p. 238). In other words, bodies are never fully autonomous. With the trans-corporeal the human is not a bounded vessel, and "travel[s] through the entangled territories of material and discursive, natural and cultural, biological and textual" (ibid). In other words, the more-than-human always already resides in the human. An example might be the proliferation of micro-plastics found in oceans, animals, and humans alike, or as Neimanis notes, the fact that "industrially produced chemicals are found in the blood and breastmilk of every single living subject" (2017, p. 40). I am thinking of the human through Alaimo's trans-corporeal lens, which itself draws upon Barad's theory of intra-action and new materialist theories of non-human agency (see Alaimo, 2018, p. 436). With the trans-corporeal, the human body is inseparable from objects, places, histories, cultures, technologies, and ecologies altering and being altered by one another in dynamic flows and interchanges. However, I also recognise, following Neimanis who looks to writer activists, Audre Lorde and Adrienne Rich amongst many other feminist, decolonial scholars, that a politics of location is necessary in recognising that while we are trans-corporeal, that environmental destruction and its consequences are not shared evenly.

The 'object of our knowledge' and the 'problem' in assemblage

"We don't obtain knowledge by standing outside the world; we know because we are of the world. We are part of the world in its differential becoming" (Barad, 2007, 185). This has effects on our teaching and learning and opens options for the use of creative practice research, especially that linked to a site, to an important form of knowledge production. Our bodies are not pockets for our brains. With creative practice research the knowledge production may not be a strict answering of questions through observation and representation. Instead, maybe the knowledge is in understanding our entanglement or our trans-corporeality with the world. An embrace of this form of embodied learning that takes place amongst the world of that which is being learned, in *Frutti Tutti*'s case, *fruit* accepts material-discursive-naturecultures and finds ways to extend knowledge and thought beyond intellectual reason alone to a trans-corporeal multiplicous connection to fruit in its multiple guises and becomings. In this sense, in editing *Frutti Tutti*, I am making agential cuts by interviewing people and editing the artwork. I create a renewed VwO, which I enfold into the site, where the participants following the map become with the market in a renewed way. They are a BwO and trans-corporeal. An assemblage of forces and intensities but not a closed whole. It is through the performativity of doing and becoming the artwork that knowledge is produced. Therefore, I set the conditions of the work, but I am not separate to it. I do not produce the knowledge, but my artistic intervention creates the scene/s and institutes the progression that comes from it. Instead, knowledge comes from the entangled relations that take place within the BwO. Referring to my previous question about how to measure the project's 'success' becomes rhetorical. It is the conditions that come into being through the interaction with the work, or in performative terms, the bringing into being of knowledge through the doing. As Lather and St. Pierre (2013) state:

> [H]ow do we determine the 'object of our knowledge' – the 'problem' we want to study in assemblage? Can we disconnect ourselves from the mangle somehow (Self) and then carefully disconnect some other small piece of the mangle (Other) long enough to study it? What ontology has enabled us to believe the world is stable so that we can do all that individuating? And at what price? How do we think a "research problem" in the imbrication of an agentic assemblage of diverse elements that are constantly intra-acting, never stable, never the same? (p. 630)

To return to my original question about impartiality in this research, I hope that I have demonstrated that I, as researcher, am never impartial but I am also not at the centre of the art outcome, and indeed the outcome in a work such as *Frutti Tutti* is unfixed. The conversations wove together alongside me. Coming to the soundscapes as recordings after the fact, participants are entangled into these conversations, where historicity takes place and the binary assumptions around humans, fruit, and place are enmeshed in an assemblage of relations. They see colour, hear the market spruikers, the exchanges of other shoppers, and smell the multiplicity of organic and otherwise. There is the human and more-than-human, but we are never separate. As a methodological approach in EE research, I am creating the space for knowledge production through the performativity that takes place. The creative work has its own agency.

All words are spoken and walked into being, entangled against one another. Biodiversity, fruit species, economics, worker exploitation, price gouging, food waste, transport, memory, family, colour, senses, culture, farming, selling, location, country, poverty, greed, giving, nurturing, and love. This is a local ecology, empirically grounded, embedded, and embodied research, rather than universal. This is a step towards a pedagogy that opens itself beyond human exceptionalism to further learning about the human and the more-than-human. It is a pedagogy that works towards ecosocial justice, decolonial approaches, and is both queer and feminist.

In their book, *Creative Agency*, Harris (2021) asks:

> What if 'knowing' were not the measure of knowledge, but instead 'listening', 'asking', 'sitting' were? What if innovation were not the measure of creativity, but rejection of the known? (p. 179)

They instead embrace a "creative wilding" that does not follow "truth claims" (p.180). Rather, they implore the researcher to connect people to stories that are bigger than their individual selves. Within EE there is a place for art practice that cannot escape its performative enactments but also doesn't try to. In recognising that certain techniques within art can produce new ways of experiencing the environment or of understanding the complexity of ecosystems or the intersections of the social, political, and personal within nature, art can produce empathetic and embodied knowledge-affect.

References

Ahmed, S. (2008) *Queer phenomenology: Orientations, objects, others*. Duke University Press E-Duke.

Alaimo, S. (2018). Trans-corporeal. In R. Braidotti & M. Hlavajova (Eds.), *Posthuman glossary* (pp. 435-438). Bloomsbury Academic.

Alaimo, S., & Hekman, S. (Eds.). (2008). *Material feminisms*. Indiana University Press.

Barad, K. (2003). Posthumanist performativity: Toward an understanding of how matter comes to matter. *Signs: Journal of Women in Culture and Society*, *28*(3), 801-831.

Barad, K. (2007). *Meeting the universe halfway: Quantum physics and the entanglement of matter and meaning*. Durham, NC: Duke University Press.

Barad, K. (2011). Nature's queer performativity. *Qui Parle*, *19*(2), 121-158.

Butler, J. (1999). *Gender trouble: Feminism and the subversion of identity*. (10th anniversary edition.). Routledge.

Centre de Cultura Contemporània de Barcelona. (2022, March 17). *Rosi Braidotti: "The concept of human has always been associated with relations of power"*. [Video]. Youtube.
Deleuze, G., & Guattari, F. (1984). *Thousand plateaus: Capitalism and schizophrenia*. University of Minnesota Press.
Haraway, D. (2016). *Staying with the trouble: Making kin in the Chthulucene*. Duke University Press.
Harris, D. (2021). *Creative Agency* (1st ed. 2021.). Springer International Publishing.
Jungnickel, K. (2020). How do you make and communicate your research? In K. Jungnickel (Ed.), *Transmissions: Critical tactics for making and communicating research.* The MIT Press.
Lather, P., & St. Pierre, E. A. (2013). Post-qualitative research, *International Journal of Qualitative Studies in Education*, *26*(6), 629-633.
Law, J., & Urry, J. (2004). Enacting the social. *Economy and Society*, *33*(3), 390-410.
Manning, E. (2020). Against method. In *The Minor Gesture* (pp. 26-45). Duke University Press.
Mazzei, L. A. (2013). A voice without organs: Interviewing in posthumanist research. *International Journal of Qualitative Studies in Education*, *26*(6), 732-740.
Neimanis, A. (2017). *Bodies of water: Posthuman feminist phenomenology*. Bloomsbury Academic, an imprint of Bloomsbury Publishing Plc.
Polanyi, M. (2009). The tacit dimension. In *Knowledge in Organisations* (pp. 135-146).
St. Pierre, E. A. (2011). Post qualitative research: The critique and the coming after. In N. K. Denzin & Y. S. Lincoln (Eds.), *Handbook of qualitative inquiry* (4th ed., pp. 611-635). Los Angeles, CA: Sage.
St. Pierre, E. A. (2015). Afterword: Troubles with embodiment. In *Methodologies of Embodiment* (1st ed., pp. 138-148). Routledge.

Cassandra Tytler (BA, BFA (Hons), MFA, PhD) - Centre for People, Place & Planet, Edith Cowan University, Western Australia.

Cassandra Tytler is a video artist and scholar situated within the Centre for People, Place, and Planet at Edith Cowan University. She works within the School of Education and Western Australian Academy of Performing Arts. Her research interests lie in the performance of video and its encounter within place, to create a relational and aware politics of resistance to normalising narratives of exclusion. Her current project, *A Whisper, A Chorus*, uses site-specific practice to amplify stories that inspire climate-

informed action, empathy, and inclusion. Its aim is to contribute to creating a sense of the 'more-than-individual' in a rhetorically compelling way. In 2023 she was awarded a Creative and Performance Leadership Fellowship with the Forrest Foundation.
https://orcid.org/0000-0003-0357-7123

CHAPTER 13

EMERGENT ENVIRONMENTAL EDUCATION INQUIRY: A METHODOLOGY OF THINKING WITH THINGS

SCOTT JUKES

Abstract

In contemporary times the world is faced with many environmental challenges. Yet old habits of human hubris persist. Our times call for us to think differently, and in the context of environmental education research, think methodology differently. This chapter offers four provocations that challenge orthodox educational research practices. Namely, challenges to conventional humanist qualitative inquiry (St. Pierre, 2011), anthropocentric data, overly prescriptive methodocentrism of educational research (Weaver & Snaza, 2017) and inquiry grounded within paradigmatic structures (Gough, 2016). These challenges confront the issue of overly dogmatic and dictatorial research methodology. This chapter takes these provocations seriously, aiming to look beyond anthropocentric, paradigmatic and overly deterministic approaches to inquiry. I propose that an approach for doing this may be to think with things, such as theory, concepts, landscapes, ecosystems, non-human species, and materiality. This is a process of assembling relations in research practice that allows for unexpected possibilities to emerge—a process that allows for responses to the significant and account for the ephemeral. As part of this approach, I embrace the notion of "follow[ing] the matter-flow" (Deleuze & Guattari, 1987, p. 479) within the context of inquiry. I offer some examples to evoke this notion of open-ended inquiry.

Key words: Thinking-with, Emergent inquiry, Assembling relations, Posthumanism, Anthropocentrism

Framing and overview

Contemporary times pose some unique challenges for education, especially environmental education. The 6th mass extinction, the 4th industrial revolution, the escalating climate crisis, fires, floods, and resultant ecocide are wreaking havoc across the planet. But these are not just issues impacting far-away places, these are issues challenging communities close to home, occurring on both local and global scales. Our predicament calls for us to think differently, to alter our unsustainable ways of relating to more-than-human worlds.

But how might we think differently? Haraway (2016) argues that "it matters what thoughts think thoughts" (p. 35). Which is to say, thought is multi-layered, with a multiplicity of cultural assumptions and worldviews at play in the ways we think. For me, drawing upon Stewart (2020), it is not just that we need to think differently, but we need to think thinking differently. Or as St. Pierre (1997)—inspired by Deleuze and Guattari—states, "to produce different knowledge and produce knowledge differently" (p. 175). Which, in the context of research inquiry, prompts me to re/consider methodology and some of the common, taken for granted, practices. After all, methodology guides how we think and approach our research. Thus, it matters what thoughts guide methodology.

Hence, this chapter is a philosophical exploration of methodology, with my field of outdoor environmental education the touchstone for many of the discussions. The question that prompts this chapter asks: how might we think and do methodology differently in environmental education research? I do not intend to definitively answer this question, but instead use the prompt in challenging some common practices. I offer these challenges as provocations to help us think harder about unquestioned practices. Namely, I will explore four provocations: i) challenges to conventional humanist qualitative inquiry (St. Pierre, 2011), ii) the idea that humanist qualitative inquiry is anthropocentric, iii) overly prescriptive methodocentrism of educational research (Weaver & Snaza, 2017), and iv) inquiry grounded within paradigmatic structures (Gough, 2016). These confront the potential issue of overly dogmatic and dictatorial research methodology, while offering some alternative possibilities along the way.

Following on from the four provocations, I will explore further opportunities for emergent inquiry practices that are made possible from the earlier provocations. Specifically, I will explore the idea of thinking with things (such as, theory, concepts, landscapes, non-human materiality). Following this idea, I will offer some examples for empirical inquiry.

Challenges to conventional humanist qualitative inquiry

As an early research student, I found myself enrolled into various research methods subjects to help me develop my understanding and approaches to research. These core subjects were generally framed around qualitative and quantitative approaches to research. From this, the focus quickly shifted to the type of data us research students wanted to source and use in our studies. Classically, it was framed as qualitative equals words and quantitative equals numbers. For myself, with an undergraduate degree that mostly had a hard scientific focus—a focus that did not resonate with how I experienced the world as an educator—I quickly angled my focus towards qualitative approaches.

As I explored the dominant textbooks and popular readings, the focus for qualitative inquiry was largely on the data collection, which discussed how we would engage with people to gather data. Surveys, interviews, participant observation etc. were the popular conventional methods I was reading about. Yet, my supervisor interjected and warned me about the "rush to application" (St. Pierre, 2016, p. 112), sending me a range of philosophical readings that touched on feminist theory, poststructuralism, new materialism and new empiricism, posthumanism etc. Although confused and a little overwhelmed as a master's student, I started to gain a different perspective and see things differently. I read St. Pierre's (2011) critique of conventional humanist qualitative inquiry and started to rethink how I might approach research… something I have continued with since.

St. Pierre (2011) argues that conventional humanist qualitative inquiry is overdetermined. The common process is to take accepted practices and follow them rigidly. By following rigid design and processes, researchers may come to an objective truth. This may be technically correct and the way for scientific inquiry, however, for qualitative inquiry this process is steeped in positivism. As St. Pierre (2011) argues, such a mode of inquiry involves structures that are tightened, disciplined, regulated, and normalised. On one hand this may sound like a good thing, however, this may lead to homogenisation where it is difficult for differences to emerge. In other words, by reducing and limiting what is 'accepted', research inquiry may become hegemonic and lose its radical potential to produce new knowledge and think differently about the world and our place in it.

Another potential pitfall of conventional approaches to research involves the drive for generalisable and universal findings. Obviously, we want our research to be useful, to do something and to have broader implications. However, a potential by product of trying to generalise and universalise findings is that we smooth out differences that appear on the margins and lose some of the contextual specificity. As St. Pierre (2011) states, "we know that the most significant data in a study might occur only once" (p. 622). I believe it is important to stay open to the possibility of something unexpected and have a reflexivity during inquiry so we may attend to the unexpected. In short, qualitative inquiry can have radical potential for creating new knowledge and new ways of thinking, yet possibly not if it becomes too constrained.

Humanist qualitative inquiry as anthropocentric: From data to empirical materials

Through St. Pierre's (2011) critique of conventional humanist qualitative inquiry, she offers a coming after: post qualitative research (which has later morphed into post qualitative inquiry). There has been a spike in interest of such ideas (e.g., see the previous edition of this book (White et al., 2022) and Hart and White (2022)). Yet Annette Gough and Noel Gough (2022) highlight that the term 'qualitative' tends to signify a mode of data production and the label post qualitative inquiry may not be an appropriate descriptor. I will touch on this more later, but for now it signals my interest in what counts as data and what is common in qualitative inquiry.

From my experience, there is a large portion of educational research that tends to focus inquiry on humans, or the humanist subject. As St. Pierre (2016) explains, "typical social science research methods and practices (e.g., experiment, interview, observation) are … dependent on the humanist subject" and "typical social science research concepts (e.g., data, research design, data analysis, measurement, researcher) are also grounded in the humanist subject" (p. 122). This practice often places humans as the grounds for knowledge production (or data production). Furthermore, it relies on the notion of the human as a bounded individual. Thus, this makes me think that conventional humanist qualitative inquiry is anthropocentric. This is problematic, especially in the context of mass extinction of species and a climate crisis, as human exceptionalism narrows a researcher's vision, ignoring and subjugating non-human actors and human relational entanglement with the world.

The notion of post qualitative inquiry is not another alternative methodology but a provocation to radically rethink and reimagine approaches to research (including what counts as data). It is not that 'conventional' data doesn't count, but instead, what else might count? What else can inform our inquiry? What else can we draw upon that is valuable, that might not be recognised as a conventional method? What are some other sources of knowledge production for (post) qualitative inquiry? What can non-humans or more-than-humans tell us in educational research? What acting role may non-human entities play in educational contexts? These are questions that spur me on in my research practice, provoking me to try creating new knowledge and to think thinking differently.

Such thinking has led me to put aside the term data and, as suggested by Denzin (2013), embrace the term empirical materials. Empirical materials are something created through practice and are used to think with and work through, rather than data that holds a fixed truth. Empirical materials can look beyond conventional humanist data (such as interviews) and, as such, move beyond an anthropocentric focus. Examples of empirical materials I have thought with include a picture story book created by a student that engages with more-than-human worlds (Jukes & Reeves, 2020), a piece of 'junk' timber used to make a canoe paddle (Jukes, 2020) and events and more-than-human encounters depicted through photos and video (Jukes, 2021; Jukes et al., 2019, 2022, 2023). Possibilities in this realm may also be conceived as arts-based methods, transgressive data (St. Pierre, 1997), attuning-with (Riley & White, 2019) material actors in educational contexts, including non-human entities. For me, the meanings of empirical materials are in motion and theory helps to read different insights related to the topic of inquiry.

Methodocentrism in educational research

The third provocation I offer is the notion of 'methodocentrism' (Weaver & Snaza, 2017). In a sense, this follows similar contours of the critiques offered above. Weaver and Snaza (2017) argue that much of conventional Western thought follows a belief in the certainty of facts: predictability, stability, and consistency. They call this the "Laplacean dream" (p. 1055)—the belief in determinism. This can lead to Western science being caught in methodocentrism; "the belief that particular, pre-formed methods can guarantee the validity of an intellectual investigation into the world by factoring out the vicissitudes of the observer's entanglement with the world" (Weaver & Snaza, 2017, p. 1056). They argue that the world is not so certain and dependable and such thought contains an anthropocentrism that "erases non-human beings" (p. 1055).

I see methodocentrism as a methodological approach where aspects of attention and measurement are worked out in advance. A practice of putting the cart before the horse, perhaps? From this, the researcher might focus too much on set methods and not be alive to the emergent aspects and more-than-human elements that can arise in the research context. Furthermore, such practices often involve the separation of subject from the object of inquiry, effacing the researcher's entanglement with the world. St. Pierre (2018) explains that:

> The experimentation required in [what she calls] post qualitative inquiry cannot be accomplished within the methodological enclosure. This experimental work is risky, creative, surprising and remarkable. It cannot be measured, predicted, controlled, systematized, formalised, described in a textbook or called forth by preexisting, approved methodological processes, methods, and practices. (p. 604)

In addition, Weaver and Snaza (2017) suggest that many pre-existing methods are inadequate for attuning to the more-than-human world. For me, an important research aim for contemporary times is to engage and pay attention to the more-than-human world. But how can we listen to what the more-than-human has to say? This involves letting go and embracing a position of not-knowing, where the more-than-human world then has a chance to tell us something. Weaver and Snaza (2017) emphasise that "educational researchers must come to invite the (previously) inaudible voices of non-humans into their discussions if we hope to have any relevance in our more-than-human world" (p. 1059). To do this, we may avoid setting too many pre-determined and anthropocentric methods in advance.

Postparadigmatic materialism as a new mode of inquiry

The last provocation I will explore is Noel Gough's (2016) suggestion that the rhetoric of paradigms is exhausted and can over structure research (also see Annette Gough & Noel Gough, 2022). As he argues, "it is commonplace for commentators on trends and issues in outdoor and environmental education research to describe methods and methodologies by reference to paradigmatic categories" (p. 52). However, reading Gough's argument points out two issues with paradigmatic framing. Firstly, the dialectic of paradigms can create a polemic where paradigms are segregated and set against one another. This works against the messy realities of scholarship whilst being adversarial, building research projects that privilege neat, tidy, and linear ways of knowing.

Secondly, Gough (2016) contends that "paradigm" is a structuralist concept and poses a "significant difficulty" as it is "predicated on an implicit model of the metaphorical equivalence of the physical and social constructions of our world(s)" (p. 53). To put this simply, paradigms are seen as the base, ground or foundation of research, from which researchers then build frameworks and other elements of inquiry. Gough asserts that such constructions of reality and research are problematic, and do not work with perspectives such as poststructuralism. Notably, this is not about another paradigm shift, but about placing a question mark against the structuring and at times oppositional notion of paradigms. From my perspective, considering paradigm as a structuralist concept reduces complexity and simplifies both reality and research. Influenced by the likes of Law (2004), I am more inclined to embrace the messiness of inquiry and leave open possibilities for emergent materialisations in my research contexts.

To offer an alternative to paradigmatic structuring, Gough (2016) suggests the term postparadigmatic materialism to describe a new movement in thought, drawing upon the material turn and posthumanist philosophies (e.g., Barad, 2007; Coole & Frost, 2010; Deleuze & Guattari, 1987). Annette Gough and Noel Gough (2022) further elaborate on such dispositions, suggesting researchers plug theory into empirical materials (drawing upon Jackson & Mazzei, 2011). Taking these suggestions, I have worked with postparadigmatic ideas in my research to challenge hierarchical anthropocentrism and engage with more-than-human worlds in attempts to think differently (for example, see Jukes & Reeves, 2020; Jukes, 2020, 2021; Jukes et al., 2019, 2022, 2023).

In summary, I have proposed four provocations that provide challenges to conventional approaches to research. These provocations have prompted me to reconsider what counts as data, rethink strict pre-determined methods and loosen rigid structures. Thus far, I have offered some alternative ideas. Next, I propose some more emergent and open-ended possibilities for research inquiry, that may help us think methodology differently.

An emergent and open-ended possibility: Thinking *with* things

> Things are their relations. (Ingold, 2011, p. 70)

In efforts to engage with different modes of inquiry for environmental education research, I have embraced process ontologies and relational ways of thinking, where a theory of life and knowledge production are entangled in practice. As the Ingold quote at the beginning of this section signals, things (or entities) are a bundle of relations. No thing acts alone, as they are bound and conditioned by a multiplicity of relations (Bennett, 2010). As Braidotti (2022) quoting Rovelli states, "the world is a continuous, restless swarming of things; a continuous coming to light and disappearance of ephemeral entities... A world of happenings, not of things" (Rovelli, 2014, quoted in Braidotti, 2021, p. 111). So, really, things are not things at all, but happenings, relations, and processes. Or, in other words, emergent potentialities. Process philosopher Bergson (1998) writes that "a slight effort of attention would reveal to me that there is no feeling, no idea, no volition which is not undergoing change every moment" (p. 1). Thus, these ideas lead me to consider a mode of inquiry that is active, moves and thinks with a multiplicity of happenings.

When we open ourselves to emergent inquiry, to use St. Pierre's (2018) words, the "focus is not on things already made but on things in the making" (p. 604). Thus, I consider such a mode of inquiry as assembling relations. I consider assembling involving thinking with multiple things together. As suggested above, I might think with theory and empirical materials to gain new insights. Nail (2017), drawing upon Deleuze and Guattari's (1987) concept of assemblage, helps me elaborate and provide an example:

> If we want to know what something is, we cannot presume that what we see is the final product nor that this product is somehow independent of the network of social and historical processes to which it is connected. For example, we cannot extract the being of a book from the vast historical conditions of the invention of an alphabetic language, distribution of paper, the printing press, literacy, and all the social contingencies that made possible the specific book of our inquiry, with all of its singular features (color, lighting, time of day, and so on) and all the conditions under which someone is inquiring into the book. A vast network of processes continues to shape the book and thus there is no final product. (p. 24)

Thus, we may never know the book in its totality, but we can inquire into the book, thinking with its material, social, political historical, and other features. This is but one example, however, it highlights the vast connectedness of a given thing, where its relations condition its existence.

As a mode of inquiry, following connections and the capacity of things to act can be linked to new materialist/relational materialist ideas. Bodén and colleagues (2019) highlight that relational materialism was "first articulated and framed within [Bruno Latour's] Actor Network Theory" (p. 1). Muecke (2022), in an essay acknowledging the importance of Latour's philosophy following his death, explains that Latour taught us to think with the things of the world. Things, whether they be human or non-human, are entities that act (or have the capacity to act) and help produce worlds. There is a pluralism inherent in such thinking that moves beyond anthropocentrism, accepting that things are stitched to and shape the world and its political networks. Following a thing as an actor imbricated in a broader ecology can reveal unexpected and emergent potentialities.

Importantly, things (or actors) include not only humans and non-human material entities but can also include concepts. As Massumi (in the foreword to Deleuze and Guattari's (1987) *A Thousand Plateau's*) explains:

> A concept is a brick. It can be used to build the courthouse of reason. Or it can be thrown through the window. What is the subject of the brick? The arm that throws it? The body connected to the arm? The brain encased in the body? The situation that brought the brain and the body to such a juncture? All and none of the above...What interests us are the circumstances... The concept... is an act. (p. xi)

This can be described as Nomad thought, which is "immersed in a changing state of things" (Massumi, 1987, p. xi). What defines this emergent mode of thought is the dynamic and open-ended positionality, that requires an engagement with a particular context and set of conditions.

My research is specifically in the field of outdoor environmental education. In my research I have focused on particular places, times, people, non-human species, materialities and contexts, with specific theories and concepts, through various empirical materials to help me inquire into different questions (e.g., for a summary of a large portion of my work thus far, see Jukes, 2022). In other words, this is a process of thinking with theory, concepts, landscapes, human, and non-human entities and the material relations that emerge or are produced. For me, thinking through relations in the context of inquiry assembles new insights, ways to think differently and reconceive relationships in the context of precarious times.

In short, I might summarise this as a practice of "follow[ing] the matter-flow" of relations (Deleuze & Guattari, 1987, p. 479).

Thinking with a river

At the beginning of this chapter, I posed the question how might we think and do methodology differently in environmental education research? Obviously, the various texts and philosophers I have cited have helped me to try and engage with different modes of thought. However, as Ingold (2011) writes:

> Why do we acknowledge our textual sources but not the ground we walk, the ever-changing skies, mountains and rivers, rocks and trees, the houses we inhabit and the tools we use, not to mention the innumerable companions, both non-human animals and fellow humans, with which and with whom we share our lives? They are constantly inspiring us, challenging us, telling us things. (Ingold, 2011, p. xii)

This provocation from Ingold resonates with me, as it challenges some modes of hierarchical anthropocentrism. In a similar vein, Bridle (2022) argues that historically humans have only valued human intelligence, encouraging us to think beyond it. Places and environments, like rivers, have their own modes of intelligence and have provoked me to rethink research in various ways. Here, I am also partly influenced by Leopold's (1949) classic essay, *Thinking like a mountain*, where to think like a mountain encourages a broader ecological view of the environment. For this section, I will take the notion of thinking with a river as a thought experiment to see what emerges.

As I consider the idea of thinking with a river, my thoughts already begin to branch out and multiply. In a way, my thoughts might look a little like Figure 13.1. Just as Deleuze and Guattari (1987) used the concept of the rhizome as a way to move beyond arborescent thought, I have previously used the image of a river to reconsider my outdoor environmental education research (see Jukes et al., 2019). So, we may think with the river as a concept, where concepts provide orientations for thought (Deleuze & Guattari, 1994). All rivers are part of a river system, or catchment, which ebbs and flows with rainfall and other weather conditions. Some parts of the catchment are ephemeral while others flow perennially. Thus, to conceive research inquiry as a river we might understand that there are ebbs and floods, eddies, main currents, as well as silent backwaters. There may also be anthropocentric utilitarian influences like deforestation (manipulating habitats and flows rates), dams that block and impede, and even impacts from introduced invasive species. What do we attend to? What do we set our focus on? Do we sit in the one place and look at a static image or do we engage with and "follow the matter-flow" (Deleuze & Guattari, 1987, p. 479)?

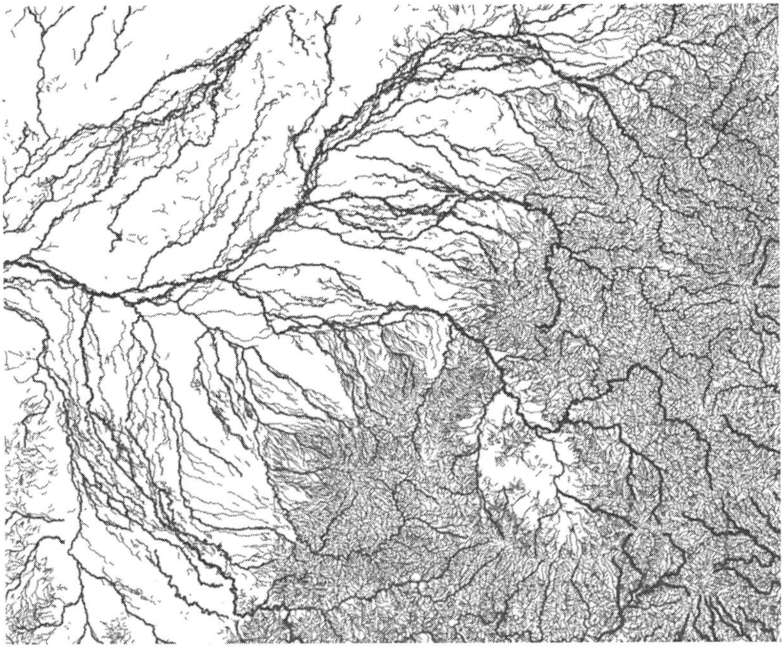

Figure 13.1: Map of part of the Murray-Darling basin created from ephemeral and perennial stream data. Image created by and reproduced with permission from Robbi Bishop-Taylor using data from Geoscience Australia under a Creative Commons Attribution 4.0 International Licence

Harold Fisk's meander maps of the Mississippi River (https://publicdomainreview.org/collection/maps-of-the-lower-mississippi-harold-fisk) offer another image of a river system that provokes my ruminations on methodology. The maps chart the multiple courses and behaviour of the river over time, visually demonstrating the way rivers constantly change, overlap, and shift. For me, staring at the images and pondering research methodology, I had the thought that the gridded farmland, with its domesticated monocultures and striated patterns of work, may reflect St. Pierre's (2011) conventional humanist qualitative inquiry. Similar patterns of use yield similar types of results. Conventional humanist qualitative research generally attends to the main currents and already domesticated grounds. Whereas the wild fluctuating river spreads, constantly making new paths, flowing with varying intensities with ephemeral events creating enduring influence.

Obviously, I am thinking with the river philosophically, however, we can also think with a river as a material reality and conduct empirical inquiries in and with such entities. Which is to say we can physically go to a river and sit with it, walk next to it or even paddle along it, and doing so can tell us something. This is a process I have explored with others elsewhere (see Jukes et al., 2022), where we examined how a riverscape shaped an outdoor environmental education river journey and the potential learning events that emerged. To do so, we utilised a range of methods such as photography and video as empirical materials, helping us respond to the ebbs and flows of the riverplace. Analysis took the form of more-than-human analysis, which looked to account for the material actors that played a part in conditioning the river environment and our journey along it. Such a methodological process was able to account for the emergent and unexpected aspects, whilst attuning us to more-than-human worlds.

For me, thinking with the river as a methodological process also requires a critical edge, where it is important to consider what constitutes it at multiple scales. This includes ecologically, politically, historically, and culturally. At micro scales this might mean specific detail, with another important step to zoom out and look at the upstream and downstream effects (both literally and figuratively). I see this as a potential contemporary approach to environmental education research, as such a methodology can inform curriculum development and approaches to pedagogical experimentation in relation to rivers and other environments.

Haecceities—becoming with the *this*ness of things

> Taking a walk is a haecceity. (Deleuze & Guattari, 1987, p. 307)
>
> Every becoming is a block of coexistence. (Deleuze & Guattari, 1987, p. 340)

To offer one more possibility for thinking with things, I will offer the concept of haecceity and share some events from a recent walk with students in the mountains of Victoria. But where to start? The Duns Scotus origins of the concept? The framing of a bushwalk with some outdoor environmental education students? I could even start 600 years ago around the speculated origin of the Snow Gum (the thing) I am thinking of, near Derrick Hut at the top of Swindlers Spur. But beginnings are always elusory and suggest linearity and boundaries, which is something I am striving to move against. I am aiming to embrace something a little messier here (messy like the crinkled bark and flailing limbs? Not sure yet.). Deleuze and Guattari (1987) write that a "haecceity has neither beginning nor end, origin nor destination; it is always in the middle. It is not made of points, only of lines. It is a rhizome" (p. 307). I guess what I am trying to capture here, in this little piece of writing, is a mood of a moment. A non-subjective singularity.

Figure 13.2: A moment with a very old tree

There were some comments heard coming from the hut. The other group were in there, looking at us hug the tree (Figure 13.2). The lads, wondering (out loud) why you would bother to hug a tree. It's just a tree, after all. A tree that has survived 600 years in conditions we endured for only one night. Bravado is in the air, oozing from the hut. My group didn't care about the bravado. They got caught in the trees orbit, its duration. There were later confrontational retorts challenging the narrowness and diversity of views. Then laughing it off. A few days later we walked past the same tree.

Someone suggested we hug the tree again, or maybe it called out to us for a cuddle. We all indulged.

> Climate, wind, season, hour are not of another nature than the things, animals or people that populate them, follow them, sleep and awaken within them. This should be read without pause: the animal-stalks-at-five-o'clock. The becoming-evening, becoming-night of an animal, blood nuptials. Five o'clock is this animal! The animal is this place! (Deleuze & Guattari, 1987, p. 306)

The snow gum endures and coexists with us. The snow gum *is this place* and we are dimensions of its multiplicity, in relation. But we are also of the other things. The comments, the bravado, the cultural ecologies that shape the various world views coming together, morphing and evolving. We are of the differing intensities too—slowing down to look at a tree, to consider its mode of existence, a life shaped by wind, snow, altitude, and climate. The carved initials at the base, someone's attempt at marking identity. The hut nearby built for refuge after a tragedy, giving some travellers a sense of the familiar through its four walls and a roof. These things coexist in a mode of individuation. Haptic phenomena shaping events.

> There is a mode of individuation very different from that of a person, subject, thing, or a substance. We reserve the name *haecceity* for it. A season, a winter, a summer, an hour, a date have a perfect individuality lacking nothing, even though this individuality is different from that of a thing or a subject. (Deleuze & Guattari, 1987, p. 304)

Or as Clarke (2017) writes:

> A haecceity is a thing defined by its *thisness*, its process of becoming, and, in contrast to the question 'what type of thing is that?' a more appropriate response might be 'look at this! What's it/they doing/producing?!' as haecceities are by definition multiplicities, each thing one *and* many, and unique in their becoming. (p. 313)

The tree hugging was an attempt to deliberately embody the environmentalist trope, and decentre ourselves a little after a cold night on the hard ground. It worked, for some. But it also produced a thisness. One grouping listened to the story of a tree and became enveloped in a life. Some others did not engage with such an idea. But what is interesting to investigate—and what I am trying to offer a vision of—is the multiple enactments going on in such an event and the many non-subjective actors beyond the tree, an educator or individual subjects. The multiple affects of that moment that held tensions and mixtures together in the "individuality of a day, a season, a year, *a life* (regardless of its duration)—a climate, a wind, a fog, a swarm, a pack (regardless of its regularity)" (Deleuze & Guattari, 1987, p. 306). This is the fabric of the world of which we were of, that we were and are still becoming with.

St. Pierre (2011) suggests haecceity as a concept that may be thought with to de-individualise and explore the "non-subjective assemblage of humans, time, space, physical objects, and everything else" (p. 618). My offering here is a small exploration into a moment. This moment can unravel much further, and so can such investigations. However, I will pause here, leaving with the proposition that thinking in such ways is open ended, and it is such generative potential that may allow new insights and perspectives to emerge.

Conclusion

Throughout this chapter I have walked through some provocations that challenge conventional approaches to qualitative research, followed by the ideas of emergent inquiry and thinking with things. I have played with the idea of thinking with a river, both as a philosophical orientation for methodology, along with the practical idea of thinking with a river in the context of outdoor environmental education research. I have also added a moment with a tree and the idea of haecceity into the mix. To round out this chapter, I once again return to the words of St. Pierre (2011), that I believe offer some useful advice for visualising emergent modes of inquiry:

> I imagine a cacophony of ideas swirling as we think about our topics with all we can muster—with words from theorists, participants, conference audiences, friends and lovers, ghosts who haunt our studies, characters in fiction and film and dreams—and with our bodies and all the other bodies and the earth and all the things and objects in our lives—the entire assemblage that is a life thinking and, and, and… All those data are set to work in our thinking, and we think, and we work our way somewhere in our thinking. My advice is to read, and analysis, whatever it is, will follow. *(Do tell me what you think you are thinking with when you think—what are your data? And do tell me what you think you do when you think—when you do analysis? Do that.)* (St. Pierre, 2011, p. 622)

Doing something different to conventional approaches means taking a risk and embracing uncertainty. However, in doing so we become responsive to the differences that emerge in the process of inquiry. What they might be we cannot be sure of in advance. As Deleuze and Guattari (1994) contend, "to think is to experiment, but experimentation is always that which is in the process of coming about—the new, remarkable, and interesting that replace the appearance of truth and are more demanding than it is" (p. 111). What I have proposed in this chapter is the idea of thinking with things as mode of inquiry. In such a process, things/happenings/actors guide us into their worlds and the relations that condition them. Concepts, theoretical rigor and a broad image of what may count as empirical materials can help us ensure we attend to matters of significance. What I believe this can do is open us to worlds we may not have considered, taking us beyond our common tracks and pathways of thought. This, maybe, could lead us to think and relate to the world in different ways.

References

Barad, K. (2007). *Meeting the universe halfway: Quantum physics and the entanglement of matter and meaning*. Duke University Press.

Bennett, J. (2010). *Vibrant matter: A political ecology of things*. Duke University Press.

Bergson, H. (1998). *Creative evolution*. Dover.

Bodén, L., Lenz Taguchi, H., Moberg, E., & Taylor, C. A. (2019). Relational Materialism. In G. Noblit (Ed.), *Oxford research encyclopedia of education* (July 2019 ed.).

Braidotti, R. (2022). *Posthuman feminism*. Polity Press.

Bridle, J. (2022). *Ways of being: Beyond human intelligence.* Allen Lane.

Clarke, D. A. G. (2017). Educating beyond the cultural and the natural: (Re)framing the limits of the possible in environmental education. In K. Malone, S. Truong, & T. Gray (Eds.), *Reimagining sustainability in precarious times* (pp. 305-319). Springer.

Coole, D., & Frost, S. (2010). Introducing the new materialisms. In D. Coole & S. Frost (Eds.), *New materialisms: Ontology, agency, and politics* (pp. 1-43). Duke University Press.

Deleuze, G., & Guattari, F. (1987). *A thousand plateaus: Capitalism and schizophrenia* (B. Massumi, Trans.). Bloomsbury.

Deleuze, G., & Guattari, F. (1994). *What is philosophy?* (H. Tomlinson, & Burchell, G., Trans.). Columbia University Press.

Denzin, N. (2013). "The Death of Data?". *Cultural Studies Critical Methodologies, 13*(4), 353-356.

Gough, A., & Gough, N. (2022). After the posts: Thinking with theory in environmental education research. *Australian Journal of Environmental Education, 38*(3-4), 388-396.

Gough, N. (2016). Postparadigmatic materialisms: A 'new movement of thought' for outdoor environmental education research? *Journal of Outdoor and Environmental Education, 19*(2), 51-65.

Haraway, D. J. (2016). *Staying with the trouble: Making kin in the Chthulucene*. Duke University Press.

Hart, P., & White, P. J. (2022). Post qualitative inquiry: Theory and practice in environmental education. *Australian Journal of Environmental Education*, *38*(3-4), 201-210.

Ingold, T. (2011). *Being alive: Essays on movement, knowledge and description*. New York: Routledge.

Jackson, A. Y., & Mazzei, L. (2011). *Thinking with theory in qualitative research: Viewing data across multiple perspectives*. Routledge.

Jukes, S. (2020). Thinking through making: Junk paddles, distant forests and pedagogical possibilities. *Environmental Education Research, 26*(12), 1746-1763.

Jukes, S. (2021). Thinking with a landscape: The Australian Alps, horses and pedagogical considerations. *Australian Journal of Environmental Education, 37*(2), 89-107.

Jukes, S. (2022). Enacting more-than-human pedagogies in response t ecological precarity: An immanent praxiography. *Australian Journal of Environmental Education*, 1-3.

Jukes, S., & Reeves, Y. (2020). More-than-human stories: Experimental co-productions in outdoor environmental education pedagogy. *Environmental Education Research, 26*(9-10), 1294-1312.

Jukes, S., Stewart, A., & Morse, M. (2019). Acknowledging the agency of a more-than-human world: Material relations on a Snowy River journey. *Journal of Outdoor and Environmental Education, 22*(2), 1-19.

Jukes, S., Stewart, A., & Morse, M. (2022). Following lines in the landscape: Playing with a posthuman pedagogy in outdoor environmental education. *Australian Journal of Environmental Education*. *38*(3-4), 345-360.

Jukes, S., Stewart, A., & Morse, M. (2023). Learning landscapes through technology and movement: Blurring boundaries for a more-than-human pedagogy. *Journal of Adventure Education and Outdoor Learning*.

Law, J. (2004). *After method: Mess in social science research*. Routledge.

Leopold, A. (1949). *A sand county almanac: And sketches here and there*. Oxford University Press.

Massumi, B. (1987). Translator's foreword: The pleasures of philosophy. In G. Deleuze & F. Guattari, *A thousand plateaus: Capitalism and schizophrenia*. Bloomsbury.

Muecke, S. (2022). The generous philosopher. https://aeon.co/essays/bruno-latour-showed-us-how-to-think-with-the-things-of-the-world

Nail, T. (2017). What is an Assemblage? *SubStance*, *46*(1), 21-37.

Riley, K., & White, P. (2019). 'Attuning-with', affect, and assemblages of relations in transdisciplinary environmental education. *Australian Journal of Environmental Education. 35*(3(, 262-272.

St. Pierre, E. (1997). Methodology in the fold and the irruption of transgressive data. *International Journal of Qualitative Studies in Education, 10*(2), 175-189.

St. Pierre, E. (2011). Post qualitative research: The critique and the coming after. In N. K. Denzin, & Y. S. Lincoln (Eds.), *The Sage handbook of qualitative research* (Vol. 4, pp. 611-625). Thousand Oaks, CA: Sage.

St. Pierre, E. (2016). The empirical and the new empiricisms. *Cultural Studies Critical Methodologies*, *16*(2), 111-124.

St. Pierre, E. (2018). Writing post qualitative inquiry. *Qualitative Inquiry. 24*(9), 603-608.

Stewart, A. (2020). *Developing place-responsive pedagogy in outdoor environmental education: A rhizomatic curriculum autobiography.* Springer Nature.

Weaver, J. A., & Snaza, N. (2017). Against methodocentrism in educational research. *Educational Philosophy and Theory, 49*(11), 1055-1065.

White, P., Tytler, R., Ferguson, J., & Clark, J. (2022). Methodological approaches to STEM education research. Volume 3. Cambridge University Press.

Scott Jukes (BExSC, MOutEnvEd, PhD) - Federation University, Institute of Education, Arts and Community, 100 Clyde Road, Berwick, 3806, Victoria, Australia.

Scott Jukes is a lecturer in Outdoor Environmental Education at Federation University, Australia. He has a love of the rivers, mountains and coastal environments of south-eastern Australia and enjoys spending time and teaching in these places. His research explores pedagogical development and methodological experimentation, where he looks to put posthumanist and new materialist theories to work in his practice. He is particularly interested in ways to challenge human exceptionalism and rethink ontological assumptions as a response to our ecological predicament.
https://orcid.org/0000-0003-3758-9413